GRANULARITY IN MATERIALS SCIENCE

Edited by **George Kyzas**
and **Athanasios C. Mitropoulos**

Granularity in Materials Science
http://dx.doi.org/10.5772/intechopen.75231
Edited by George Kyzas and Athanasios C. Mitropoulos

Contributors

Anna Angela Barba, Veronica De Simone, Diego Caccavo, Annalisa Dalmoro, Gaetano Lamberti, Matteo D'Amore, George Z. Kyzas, Athanasios C. Mitropoulos, Linchong Huang, Shuai Huang, Yu Liang, Leonardo Schippa, Guodong Liu

Notice

Statements and opinions expressed in the chapters are these of the individual contributors and not necessarily those of the editors or publisher. No responsibility is accepted for the accuracy of information contained in the published chapters. The publisher assumes no responsibility for any damage or injury to persons or property arising out of the use of any materials, instructions, methods or ideas contained in the book.

First published in London, United Kingdom, 2018 by IntechOpen
IntechOpen is the global imprint of INTECHOPEN LIMITED, registered in England and Wales, registration number: 11086078, The Shard, 25th floor, 32 London Bridge Street
London, SE19SG – United Kingdom
Printed in Croatia

British Library Cataloguing-in-Publication Data
A catalogue record for this book is available from the British Library

Additional hard copies can be obtained from orders@intechopen.com

Granularity in Materials Science, Edited by George Kyzas and Athanasios C. Mitropoulos
p. cm.
Print ISBN 978-1-78984-307-1
Online ISBN 978-1-78984-308-8

Meet the editors

Dr. George Z. Kyzas was born in Drama (Greece) and obtained his BSc (Chemistry), MSc, and PhD (Chemical Technology-Materials Science) degrees at Aristotle University of Thessaloniki (Greece). His current interests include the synthesis of various adsorbent materials for the treatment of wastewaters (dyes, heavy metals, pharmaceuticals, phenols, etc.). He has published a significant number of scientific papers (over 90), books (as author and/or editor), chapters in books, teaching notes, and reports. He also acted as guest editor in special issues of journals and presented many works at international conferences. He has been awarded with honors, grants, and fellowships for his research career/profile by the Research Committee of Aristotle University of Thessaloniki, National State Scholarships Foundation of Greece, and Stavros Niarchos Foundation.

Prof. A.Ch. Mitropoulos was born in Athens in 1957. He studied Chemistry at the University of Thessaloniki (BSc) and Physical Chemistry at the University of Bristol (MSc, PhD). In 1998, he was appointed Professor in the Department of Petroleum Engineering at Eastern Macedonia and Thrace Institute of Technology. Since 2008, Prof. Mitropoulos has been the president of the same institute. He specializes in the characterization of porous media, nanoporous materials, and membranes with in-situ techniques of adsorption and small angle X-ray scattering. He has published more than 100 journal papers, book chapters, and patents. Prof. Mitropoulos is a member of the Society of Petroleum Engineers.

Contents

Preface

Granular Materials describes many themes in the granularity of materials. Granular materials are very simple: they are large conglomerations of discrete macroscopic particles. If they are non-cohesive, then the forces between them are essentially only repulsive, so that the shape of the material is determined by external boundaries and gravity. Yet, despite this seeming simplicity, granular materials behave differently from any of the other standard and familiar forms of matter—solids, liquids, or gases—and should therefore be considered an additional state of matter in their own right. We will see that at the root of this unique status are three important aspects: the existence of static friction, the fact that temperature is effectively zero, and, for moving grains, the inelastic nature of their collisions. No one can seriously doubt that granular materials, of which sand is but one example, are ubiquitous in our daily lives. They play an important role in many of our industries, such as mining, agriculture, civil engineering, and pharmaceutical manufacturing. They clearly are also important for geological processes where landslides and erosion and, on a larger scale, plate tectonics determine much of the morphology of the Earth. Practically everything that we eat started out in granular form. All the above clearly show the importance of granular materials and indicate that the "world" of granular materials has various segments. Therefore, the target of this book is wide. Specialists, researchers, and professors from different countries have published their research into the granular materials field in this book and we are grateful for their tremendous expertise. We also wish to acknowledge the outstanding support from Ms. Kristina Kardum, Author Service Manager at IntechOpen, who collaborated tirelessly in crafting this book.

The future of granular materials is indeed bright!

Dr. George Z. Kyzas (MSc, PhD) and Prof. Athanasios C. Mitropoulos (MSc, PhD)
Hephaestus Advanced Laboratory
Eastern Macedonia and Thrace Institute of Technology
Kavala, Greece

Chapter 1

Introductory Chapter: Granularity in Adsorption

George Z. Kyzas and Athanasios C. Mitropoulos

Additional information is available at the end of the chapter

http://dx.doi.org/10.5772/intechopen.81554

1. General aspects

Nowadays, a very promising technique for desalination and generally water purification is considered to be adsorption. Various classes of pollutants can be removed with adsorption process as dyes, heavy metal ions, organic molecules, and odors. Numerous adsorbent materials were synthesized having as major target the (possible) high adsorption capacity. Complex materials, organic (or polymeric) materials, and low-cost materials are some basic types of adsorbents used for water/wastewater purification. However, the "king" of the adsorbent materials is still the activated carbon. Activated carbon is a very strong candidate for adsorption applications due to its high porosity and large surface area for the majority of possible contaminants for removal. The two main types of activated carbon used in water treatment applications are granular activated carbon (GAC) and powdered activated carbon (PAC).

One of the basic advantages of adsorbent materials is the different shape/form that they can be produced; powders, microspheres/beads, granular particles, and monoliths are some important forms of adsorbents. However, special attention can be given to granularity of adsorbent materials given the wide use of this form not only in adsorption but also in many processes. The reason about granularity can be easily given taking into mind some examples of reality. Granular materials are characterized as simple materials with an increasing number of conglomerations of discrete macroscopic particles. However, if those materials do not present enough cohesivity, then only repulsive (strong) forces among them can exist, and the final shape of them is governed by (only) external boundaries and gravity. Granular materials do not behave similarly and present many differences from one material to another (of even familiar form of matter). All above indicate a basic triple concept: static friction, zero temperature, and the inelastic nature of their collisions (for moving grains). Granular materials play an important role in many of our industries, such as mining, agriculture, civil engineering, and pharmaceutical manufacturing. Also, they have a big impact on geological processes and

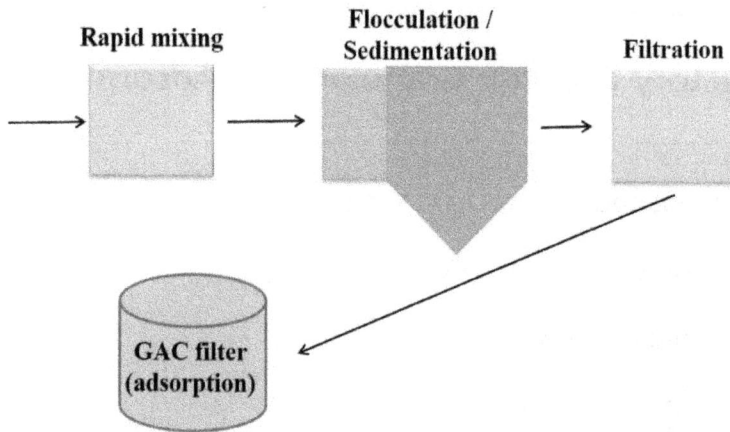

Figure 1. Post-filtration process for water purification using GAC.

erosion phenomena. So, the theory that everything eatable started out in a granular form can be easily supported and accepted. All above clearly show the importance of granular materials.

Taking into in consideration the importance of granular materials, researchers expertized in adsorption turn their interest to granular forms of activated carbons to treat and purify water (or wastewaters). From practical experience in areas where granular activated carbon (GAC) is used for drinking water treatment, it is clear that high levels of organic material in the source water result in a greatly diminished adsorption capacity and therefore a reduced lifetime of the carbon filters. It is a fact that these molecules interfere with the adsorption of other compounds present in drinking water (pesticides, taste- and odor-causing compounds, and other industrial micro-pollutants). However, few investigations have concentrated on the effect of adsorbed naturally occurring organic material on the surface properties of the carbon. The most common option for locating a GAC treatment unit in water treatment plants is post-filtration adsorption, where the GAC unit is located after the conventional filtration process (post-filter contactors or adsorbers) (**Figure 1**).

All above clearly indicated the use of granularity on adsorption technology of real industrial processes apart from the already widely known applications in engineering (mechanical, civil, electrical, etc.).

Author details

George Z. Kyzas* and Athanasios C. Mitropoulos

*Address all correspondence to: kyzas@teiemt.gr

Hephaestus Advanced Laboratory, Eastern Macedonia and Thrace Institute of Technology, Kavala, Greece

Application of the Two-Fluid Model with Kinetic Theory of Granular Flow in Liquid–Solid Fluidized Beds

Guodong Liu

Additional information is available at the end of the chapter

http://dx.doi.org/10.5772/intechopen.79696

Abstract

Numerical simulation of voidage distributions and bed expansions is carried out in a liquid–solid fluidized bed in the present work. Effects of drag force models as well as virtual mass force and lift force are studied in the prediction of particle flow characteristics; simulated results indicated that both virtual mass force and lift force could not be neglected in liquid–solid fluidized bed. Different superficial velocities of liquid phase are also studied to investigate the effects of operating conditions on the distribution of particle concentration and velocities. The coefficient of restitution varied from 0.6 to 0.99, and the effects of radial distribution function models on granular pressure and granular temperature are also studied. Different drag models exhibit various particle velocity distributions, while the Gibilaro drag model failed in predicting the liquid–solid drag to some extent in this study. A comprehensive simulation model was proposed for predicting the two-phase flow characteristics in the liquid–solid fluidized bed. Predicted axial void fraction agrees qualitatively and quantitatively well with the experimental results in the literature.

Keywords: liquid–solid fluidized bed, kinetic theory of granular flow, Euler–Euler two-fluid model

1. Introduction

The liquid–solid fluidized bed reactors are widely used in the pharmaceutical, chemical, food, petroleum, and many other industries; as a result, they have been the focus of much research. Liquid–solid fluidized beds exhibit a great variety of complex inhomogeneous flow structures. The origin and hierarchy of these of structures in liquid–solid fluidized beds have presented a challenge for both experimental and numerical research. A general understanding of their hydrodynamic behaviour is still under pursuit. One of the issues that make the accurate prediction of particulate flows to be difficult in liquid–solid flow system is the lack of accurate comprehensive simulation models and corresponded parameters.

Computational fluid dynamics (CFD) method can provide a series of fluid hydrodynamic information, which can hardly be obtained by modern measuring instruments. In the CFD modelling, the two-fluid model (TFM) assumes the liquid phase and solid phase as both continua and fully interpenetrating within each other. Among various attempts to formulate particulate flow stresses, the kinetic theory of granular flow (KTGF) is usually employed [1], which is an extension of the classical kinetic theory of gases to dense particulate flows. In this theory, the fluctuation energy of particles was described by introducing the concept of granular temperature. Thus, the particle flow behaviour can be predicted by TFM-KTGF model. A number of studies had shown the capability of the KTGF approach for modelling fluidized beds [2–7].

In the TFM, the momentum transfer between fluid and particle phases is of the great significance for the momentum of both phases. An accurate closure law for fluid-particle interactions is highly required. Generally, the interaction terms in liquid–solid flow system include the drag force, the virtual mass force and the history force except that the pressure gradient and the gravity force. The momentum exchange is mainly represented by the drag force [8]. Hence, the drag force models are important in simulating the interphase momentum transfer between the liquid and solid phases. Traditionally, the drag force models are average-based in the literature [9, 10]. With the development of computational methods and instruments, numerous CFD models were applied to the simulation of dynamic processes in the liquid–solid circulating fluidized beds. Roy and Dudukovic [11] used TFM model combining with the KTGF to simulate the flow behaviours in liquid–solid circulating fluidized beds. Razzak [12] employed the KTGF based on TFM and simulated the particle viscosity and particle pressure, and a drag model proposed by Wen and Yu was adopted to calculate the interphase momentum exchange. Cheng and Zhu [13, 14] made a comprehensive study on the modelling and simulation of hydrodynamics in liquid–solid circulating fluidized beds using both similitude method and CFD technique.

Some other studies took the local inhomogeneity of the liquid–solid flow in a circulating fluidized bed into account to calculate the interphase momentum exchange. Liu et al. [15–17] proposed a multi-scale drag coefficient model that can show better capability of predicting distribution of particle concentration. Xie et al. [18] proposed a series of correlations of KTGF model for liquid–solid flow by calculating solid pressure and viscosity and found that the particle-particle interactions can affect suspension characteristics for large particle size and high solid loading systems. Rahaman et al. [19] tested and validated three established empirical drag law correlations used to explain momentum exchange between solid and liquid phases. It was found that Wen and Yu [9] and Gidaspow [1] drag law models showed greater predictive power in terms of pressure drop and voidage in the fluidized beds of multi-particle systems. Ozel et al. [20] compared the direct numerical simulation of a liquid–solid fluidized bed with experimental data and found that fluid velocity fluctuations were mainly driven by fluid–particle wake interactions (pseudo-turbulence) whereas the particle velocity fluctuations derive essentially from the large-scale flow motion (recirculation). However, there is still an absence of comprehensive evaluation of such models for a better numerical simulation.

In present study, a comprehensive investigation of the models, kinetic theory constitutive parameters as well as models are performed; their validity in predicting the liquid–solid fluidization is compared and evaluated.

2. Liquid–solid two-fluid model

In the present work, an Eulerian multi-fluid model, which considers the conservation of mass and momentum for the solid and liquid phases, has been adopted. The kinetic theory of granular flow, which considers the conservation of solid fluctuation energy, has been used for closure. Conservation equations of mass and momentum of both phases result from the statistical average of instantaneous local transport equations. The governing equations are given below.

2.1. Governing equations

Both phases are continuous assuming a single liquid phase and single solid phase. The continuity and the momentum balance for both the phases are given. Interphase momentum transfer term includes the drag force, virtual mass force and lift force.

For simplification, we assume that (1) both liquid and solid phases are assumed to be isothermal; it is also assumed that there is no interphase mass transfer and (2) the solid phase is characterised by a spherical configuration with mean particle diameter and density. The continuity for phase i ($i = l$ for liquid phase; $i = s$ for solid phase):

$$\frac{\partial}{\partial t}(\varepsilon_i \rho_i) + \frac{\partial}{\partial x}(\varepsilon_i \rho_i u_i) = 0 \tag{1}$$

The momentum balance for the liquid phase is expressed by the Navier–Stokes equation, such equation is modified to include the drag force, virtual mass force and lift force to consider the momentum transfer between phases.

$$\frac{\partial}{\partial t}(\varepsilon_l \rho_l u_l) + \nabla \cdot (\varepsilon_l \rho_l u_l u_l) = \varepsilon_l \nabla \cdot \overline{\tau}_l + \varepsilon_l \rho_l g - \varepsilon_l \nabla p - \beta(u_l - u_s) - F_{vm} - F_{lf} \tag{2}$$

The stress tensor of liquid phase can be represented as:

$$\overline{\tau}_l = \mu_f \left[\nabla u_l + (\nabla u_l)^T \right] - \frac{2}{3}\mu_f (\nabla \cdot u_l)\overline{I} \tag{3}$$

where μ_f is combined with the laminar part and turbulent part and could be expressed as $\mu_f = \mu_l + \mu_t$. The turbulent viscosity for the liquid phase is calculated as $\mu_t = c_\mu \rho_l k^2 / \varepsilon$. The liquid phase is described by a standard $k - \varepsilon$ turbulence model.

The solid phase momentum balance is given as follows:

$$\frac{\partial}{\partial t}(\varepsilon_s \rho_s u_s) + \nabla \cdot (\varepsilon_s \rho_s u_s u_s) = -\varepsilon_s \nabla p - \nabla p_s + \nabla \cdot \overline{\tau}_s + \varepsilon_s \rho_s g + \beta(u_l - u_s) + F_{vm} + F_{lf} \qquad (4)$$

The solid phase stress tensor can be expressed in terms of the bulk solid viscosity ξ_s, and shear solid viscosity μ_s

$$\overline{\tau}_s = \mu_s \left\{ \left[\nabla u_s + (\nabla u_s)^T \right] - \frac{2}{3}(\nabla \cdot u_s)\overline{I} \right\} + \xi_s \nabla \cdot u_s \overline{I} \qquad (5)$$

2.2. Constitutive correlations

Analogical to the thermodynamic temperature for gases, the granular temperature $\theta = C^2/3$ was introduced as a measure for the energy of the fluctuating velocity of the particles [21]. The equation of conservation of solids fluctuating energy can be expressed as:

$$\frac{3}{2}\left[\frac{\partial}{\partial t}(\varepsilon_s \rho_s \theta) + \nabla \cdot (\varepsilon_s \rho_s \theta)u_s\right] = (-\nabla p_s \overline{I} + \overline{\tau}_s) : \nabla u_s + \nabla \cdot (k_s \nabla \theta) - \gamma_s - 3\beta\theta + D_{ls} \qquad (6)$$

Its description is based on the kinetic theory of granular flow, where both the kinetic and the collisional influence are taken into account. The particle pressure can be calculated as follows:

$$p_s = \varepsilon_s \rho_s \theta \left[\frac{1}{1 + \lambda/L} + 2\varepsilon_s g_0(1 + e) \right] \qquad (7)$$

g_0 is the radial distribution function at contact; it is employed to describe how density varies as a function of distance from a reference particle, which is a correlation factor that modifies the probability of collisions between grains when the solid granular phase becomes dense and can be regarded as a measure for the probability of inter-particle contact. Lun et al. [21] used the following equation for calculating radial distribution function.

$$g_0 = \left[1 - (\varepsilon_s/\varepsilon_{s,\max})^{1/3}\right]^{-1} \qquad (8)$$

In this work, the equation proposed by Bagnold [22] is used, where $\varepsilon_{s,\max}$ is the maximum particle concentration at random packing.

Arastoopour et al. [23] also derived the similar forms of equation for calculating granular pressure.

$$g_0 = \frac{1}{\left(1 - \frac{\varepsilon_s}{\varepsilon_{s,\max}}\right)} + \frac{3}{2}d_l \sum_{k=1}^{N}\frac{\varepsilon_k}{d_k} \qquad (9)$$

Ma and Ahmadi [24, 25] derived another form of radial distribution function; such form of the function is similar to that of the Arastoopour.

$$g_0 = \frac{1 + 2.5\varepsilon_s + 4.59\varepsilon_s^2 + 4.52\varepsilon_s^3}{\left(1 - \left(\frac{\varepsilon_s}{\varepsilon_{s,\max}}\right)^3\right)^{0.687}} + \frac{1}{2}d_l \sum_{k=1}^{N}\frac{\varepsilon_k}{d_k} \qquad (10)$$

Lebowitz [26] derived the radial distribution function at contact for a mixture of hard spheres:

$$g_0 = \frac{1}{(1-\varepsilon_s)} + \frac{3\sum\limits_{k=1}^{N}\frac{\varepsilon_k}{d_k}}{(1-\varepsilon_s)^2(d_l + d_k)}d_k d_l \tag{11}$$

The shear viscosity μ_s accounts for the tangential forces. It is capable of combining different inter-particle forces and using a momentum balance similar to that of a true continuous fluid. It is composed of three parts: $\mu_{s,col}$ represents the viscosity induced from the particle collisions, $\mu_{s,kin}$ represents the viscosity induced from particle fluctuations, and $\mu_{s,fr}$ from particle frictions; thus, solid viscosity can be expressed as:

$$\mu_s = \mu_{s,col} + \mu_{s,kin} + \mu_{s,fr} \tag{12}$$

$$\mu_{s,col} = \frac{5\rho_s d_s}{96\varepsilon_s}\sqrt{\frac{\theta}{\pi}}\left[\left(\frac{8\varepsilon_s}{5(2-\eta)}\right)\left(1+\frac{8}{5}\eta(3\eta-2)\varepsilon_s g_0\right) + \frac{768}{25\pi}\eta\varepsilon_s^2 g_0\right] \tag{13}$$

$$\mu_{s,kin} = \frac{5\rho_s d_s}{96\varepsilon_s\eta(2-\eta)g_0}\sqrt{\frac{\theta}{\pi}}\left(\frac{1}{1+\lambda/L}\right)\left[1+\frac{8}{5}\eta(3\eta-2)\varepsilon_s g_0\right] \tag{14}$$

$$\mu_{s,fr} = \frac{P_s\sin\phi}{2\sqrt{I_{2D}}} \tag{15}$$

where,

$$\eta = (1+e)/2 \tag{16}$$

The bulk viscosity ξ_s formulates the resistance of solid particles to compression and expansion. The following equation given by Ding and Gidaspow [2] is used in this work:

$$\xi_s = \frac{4}{3}\varepsilon_s^2\rho_s d_s g_0(1+e)\sqrt{\theta/\pi} \tag{17}$$

For the conductivity of granular energy k_s, following correlation is used:

$$k_s = \frac{25\pi\rho_s d_s\sqrt{\theta}}{128}\left[\left(\frac{1}{1+\lambda/L\eta g_0} + \frac{8}{5}\right)\left(\frac{1+\frac{12}{5}\eta^2(4\eta-3)\varepsilon_s g_0}{41-33\eta}\right) + \frac{512\eta\varepsilon_s^2 g_0}{25\pi}\right] \tag{18}$$

The rate of dissipation of fluctuation kinetic energy due to particle collisions is expressed as,

$$\gamma_s = 3(1-e^2)\varepsilon_s^2\rho_s g_0\theta\left(\frac{4}{d_s}\sqrt{\frac{\theta}{\pi}} - \nabla\cdot u_s\right) \tag{19}$$

The last term D_{ls} in Eq. (6) is the rate of energy dissipation per unit volume caused by the transfer of liquid phase fluctuations to the particle phase fluctuations. In this study, the value of D_{ls} is calculated with Koch's expression [27] as follows:

$$D_{ls} = \frac{d_s \rho_s}{4\sqrt{\pi}\theta_{g_o}} \left(\frac{18\mu_l}{d^2 \rho_s}\right)^2 |u_l - u_s|^2 \tag{20}$$

2.3. Drag model

The inter-phase drag term in the liquid and solid phase momentum equations is expressed as $\beta(u_l - u_s)$, the product of the interphase momentum exchange coefficient β and the slip velocity (relative velocity of the solid phase to the liquid phase). The inter-phase drag coefficient β can be expressed by the correlations given by Gidaspow [1]. This correlation is a combination of the works of Ergun [28] and Wen and Yu [9]; the formulation presented by Ergun [28] is used at the porosity less than and equal to 0.8 where the suspension is dense, whereas the formulation provided by Wen and Yu [9] is used for the porosity greater than 0.8, where the suspension is regarded as dilute.

$$\beta_{Ergun} = 150\frac{(1 - \varepsilon_l)^2 \mu_l}{(\varepsilon_l d_s)^2} + 1.75\frac{\rho_l(1 - \varepsilon_l)|u_l - u_s|}{\varepsilon_l d_s} \qquad \varepsilon_l \leq 0.8 \tag{21}$$

$$\beta_{Wen\&Yu} = \frac{3}{4}C_d\frac{\rho_l(1 - \varepsilon_l)|u_l - u_s|}{d_s}\varepsilon_l^{-2.65} \qquad \varepsilon_l > 0.8 \tag{22}$$

$$C_d = \begin{cases} \dfrac{24}{Re}(1 + 0.15Re^{0.687}) & Re \leq 1000 \\[2mm] 0.44 & Re > 1000 \end{cases} \tag{23}$$

However, the transition proposed by Gidaspow [1] results in the drag law discontinuous in solid concentration even if it is continuous in Reynolds number. As a matter of fact, the drag force is a continuous function for both Reynolds number and solid concentration, the abrupt change in drag at bed voidage equals to 0.8 and can also cause numerical instabilities [29], and therefore the continuous forms of the drag law is in demand to correctly simulate liquid–solid flows.

To avoid discontinuity of these two correlations, a switch function φ is introduced by Huilin and Gidaspow [7] to give a smooth transition from the dilute regime to the dense regime.

$$\varphi = \frac{\arctan[150 \times 1.75(0.2 - \varepsilon_s)]}{\pi} + 0.5 \tag{24}$$

Thus, the interphase momentum transfer coefficient becomes

$$\beta = (1 - \varphi)\beta_{Ergun} + \varphi\beta_{Wen\&Yu} \tag{25}$$

Syamlal et al. [10] proposed a correlation to calculate the momentum transfer coefficient based on the experimental data of particle sedimentation from Garside and Al-Dibouni [30]; the following equation is usually adopted:

$$\beta = \frac{3}{4}\frac{(1-\varepsilon_l)\varepsilon_l\rho_l|u_l - u_p|}{V_r^2 d_p}\left(0.63 + 4.8\sqrt{\frac{V_r}{Re_p}}\right)^2 \tag{26}$$

2.4. Virtual mass force and lift force

In gas fluidization, virtual mass force and lift force could be neglected with respect to the much smaller density of gas phase, compared with solid phase. However, in liquid–solid fluidization, these forces could not be ignored due to the fact that liquid density is usually in the same order of magnitudes as the solids.

The virtual mass force can be expressed with Ishii and Mishima [31] correlation:

$$F_{vm} = \varepsilon_s\rho_l C_{V,d}\left(\frac{du_l}{dt} - \frac{du_s}{dt}\right) \tag{27}$$

where $C_{V,d} = 0.5$.

And Drew et al. [32] expression is used to calculate the lift force:

$$F_{lf} = \varepsilon_s\rho_l C_{L,d}(u_s - u_l) \times (\nabla \times u_l) \tag{28}$$

where $C_{L,d} = 0.5$.

2.5. Boundary conditions

The aforementioned governing equations are numerically solved with appropriate boundary and initial conditions. Initially, solid particles are packed in the bed with a fixed solid concentration, and there are no motions for both the liquid and solid phases in the bed. At the inlet, the liquid velocity is constant with the concentration of unity. The particle velocity and the granular temperature are set to be zero. At the top of the bed, Neumann boundary conditions are applied to both the liquid and solid phases, and the liquid pressure is 1 atm.

The no-slip condition is set to the liquid phase at the wall. For the solid phase, the normal velocity is also set to be zero. The following boundary equations are applied for the tangential velocity and granular temperature of solid particles at the wall [33]:

$$u_{t,w} = -\frac{6\mu_s\varepsilon_{s,\max}}{\pi\rho_s\varepsilon_s g_0\sqrt{3\theta}}\frac{\partial u_{s,w}}{\partial n} \tag{29}$$

$$\theta_w = -\frac{k_{s\theta}\theta}{e_w}\frac{\partial\theta_w}{\partial n} + \frac{\sqrt{3}\pi\rho_s\varepsilon_s u_s g_0\theta^{3/2}}{6\varepsilon_{s,\max}e_w} \tag{30}$$

This simulation is carried out with the CFD codes and incorporates with kinetic theory of granular flow. To solve difference equations obtained from the differential equations, the second-order Total Variation Diminishing method (TVD) scheme is used. In all simulations, the time step is set to be constant of 1.0×10^{-3} s, such time step have been validated to be

Parameters	Value	Parameters	Value
Bed height (m)	2.04	Particle diameter (m)	3×10^{-3}
Bed width (m)	5.1×10^{-2}	Particle density (kg/m^3)	2500
Initial bed height (m)	0.35	Terminal velocity (m/s)	0.318
Superficial liquid velocity (m/s)	0.093, 0.175, 0.247	Restitution coefficient of particles	0.99, 0.9, 0.8, 0.6
Maximum packing concentration	0.63	Restitution coefficient of wall	0.99, 0.9, 0.8, 0.6
Liquid viscosity (Pa-s)	1.0×10^{-3}	Liquid density (kg/m^3)	998

Table 1. Numerical simulation parameters.

suitable for the simulation of liquid solid flows in such a fluidization system. Time-average distributions of flow variables are computed covering a period of 100 s corresponding to 2–3 weeks of computational time on a personal computer, and the last 40 s results are selected to make the average.

The solid and liquid phases treated as fully interpenetrating continua based on the extended granular flow theory; thus, the TFM is used to simulate the two-phase flows by FLUENT 6.3 based on the finite volume method. Since three-dimensional simulations are currently out of reach practically with the consideration of computation power, the simulations are carried out in two-dimensional rectangular Cartesian coordinates by ignoring front and rear wall effects.

Detailed parameter values for the simulation as well as in the experiments are reported in **Table 1**.

3. Results and discussion

3.1. Effect of drag force models

Figure 1 compares the volume fraction and the axial solid velocity against dimensionless radial position, using different drag models proposed by Gibilaro et al. [34], Huilin Gidaspow [7], Symalal-Obrien [10], and Wen-Yu [9], respectively. Experimental data are reported by Limtrakul et al. [35] From **Figure 1**, we can see that simulations by using Huilin-Gidaspow [7] model and Syamlal-Obrien [10] model are consistent with experimental data in axial solid velocity. The others deviate too much from the experimental data. Comparing the volume fraction between Huilin-Gidaspow [7] model and Syamlal-Obrien [10] model, the latter are much more different from experimental data than the former at the dimensionless radial position between 0.6 and 0.9. Hence, we chose Huilin-Gidaspow [7] drag model for further study.

3.2. Effect of virtual mass

Figure 2 shows various volume fraction against dimensionless radial position at three different bed heights of $H = 0.27$, 0.43 and 0.60 m. In this simulation, the superficial liquid velocity is

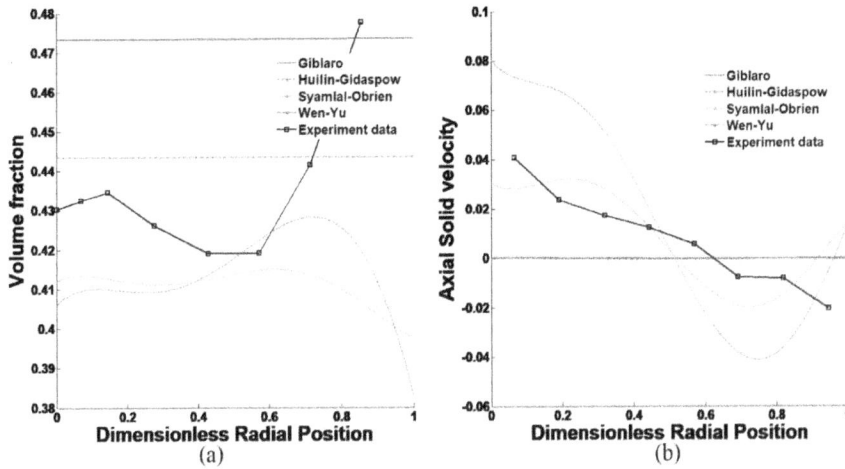

Figure 1. Influence of different drag models on (a) volume fraction; (b) axial Solid Velocity at a superficial liquid velocity of 0.07 m/s.

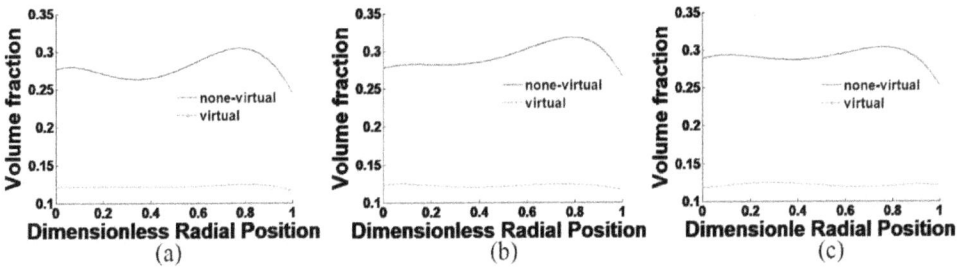

Figure 2. Influence of virtual mass on volume fraction at three different bed heights of (a) H = 0.27 m; (b) H = 0.43 m; (c) H = 0.60 m at a superficial liquid velocity of 0.10 m/s.

0.1 m/s. It is obvious that volume fraction without virtual mass is higher than volume fraction with virtual mass in the whole column. **Figure 3** shows various axial solid velocity against dimensionless radial position at three different bed heights of H = 0.27, 0.43 and 0.60 m. It is significant to notice that axial solid velocity without virtual mass is higher than solid velocity with virtual mass in the central region, lower near the wall. When the particle moves with accelerated motion in the fluid, it will accelerate the surrounding fluid in return. Due to inertia of fluid, fluid will give particle a reactive force. At this time, the reactive force will be greater than inertia force of particle itself, as if particle quality is increased, the bed voidage will also increase. This explains why volume fraction with virtual mass has a lower value. From **Figure 3**, we can see that particles go up in the central region and fall down close to the wall. Reactive forces from particles hinder the movements of fluid at the centre and reinforce the movements near the wall. So the value of axial solid velocity is lower at the central region and higher close to the wall when considering virtual mass.

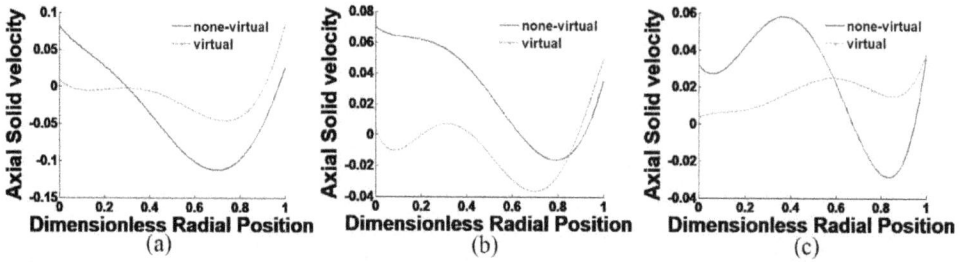

Figure 3. Influence of virtual mass on axial solid velocity at three different bed heights of (a) H = 0.27 m; (b) H = 0.43 m; (c) H = 0.60 m at a superficial liquid velocity of 0.10 m/s.

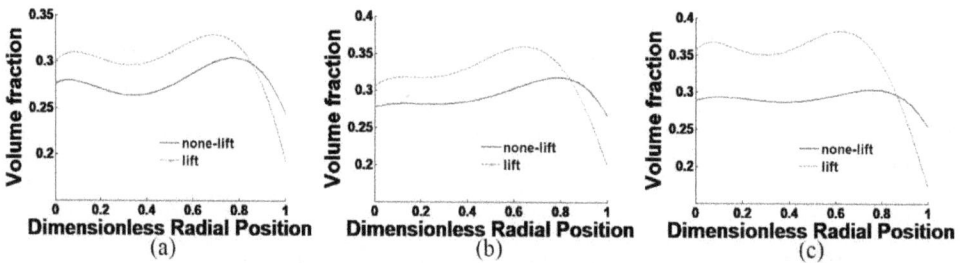

Figure 4. Influence of lift on volume fraction at three different bed heights of (a) H = 0.27 m; (b) H = 0.43 m; (c) H = 0.60 m at a superficial liquid velocity of 0.10 m/s.

3.3. Effect of lift

Figure 4 shows various volume fractions against dimensionless radial position at three different bed heights of H = 0.27, 0.43 and 0.60 m at a superficial liquid velocity is 0.1 m/s. In the simulation, lift coefficient between particles and liquid is 0.5. From **Figure 7**, it is clear that volume fraction with lift has a little higher of volume fraction than that without lift at the central region. The phenomenon is opposite near the wall, where the volume fraction of particles is lower when the lift force is considered. In general, solid volume fractions in both conditions have not much difference, not like the effect of virtual mass, and they have the similar trend. From **Figure 5**, we can see that axial solid velocity with lift is almost higher than the axial solid velocity without lift in the whole radial direction at bed height of 0.27 m. And axial solid velocity without lift is higher than solid velocity with virtual mass in the central region, lower near the wall at bed heights of 0.43 and 0.60 m.

3.4. Effect of liquid velocity

We can see that the volume fraction decrease with the increase in superficial liquid velocity obviously in **Figure 6(a)**. This is because at higher liquid velocity, it will enhance bed expansion and the bed voidage will increase, leading to decreased volume fraction. The **Figure 6(b)** shows that axial solid velocity increases with the superficial liquid velocity changing from 0.07 to 0.10 m/s while the axial solid velocity at 0.13 m/s is lower than the above two velocities.

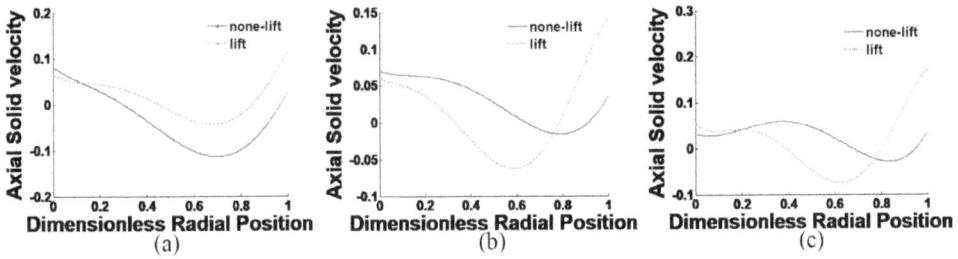

Figure 5. Influence of lift on axial solid velocity at three different bed heights of (a) H = 0.27 m; (b) H = 0.43 m; (c) H = 0.60 m at a superficial liquid velocity of 0.10 m/s.

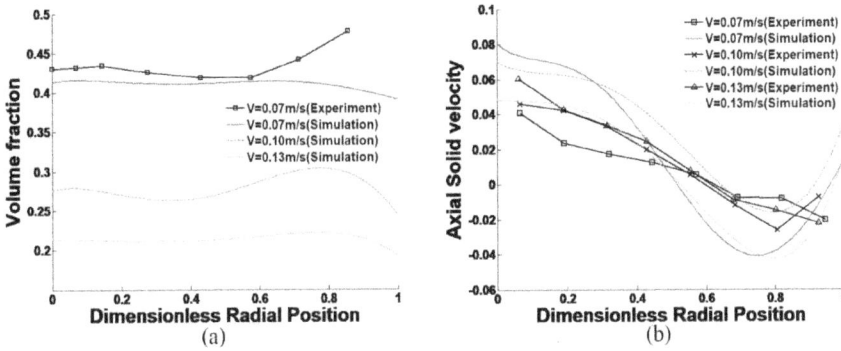

Figure 6. Influence of superficial liquid velocity on (a) volume fraction and (b) axial solid velocity.

Normally, the increase in superficial liquid velocity increases the energy input to the system, leading to enhanced solid motion. The deviation of axial solid velocity at 0.13 m/s may be due to serious fluctuation of energy at a high velocity.

3.5. Comparison of wall effects

In this work, the Johnson and Jackson [33] wall boundary condition with different specularity coefficients are used. The volume fraction and axial solid velocity are plotted in dimensionless radial position at different bed heights of $H = 0.27$, 0.43 and 0.60 m above the inlet. In **Figure 7**, between the centre and wall (about at the dimensionless radial position of 0.8), volume fraction reaches its maximum. The volume fractions are comparatively close to each other for all values of e at central region. The volume fractions increase at central region and decrease near the wall from 0.8 to 1. From **Figure 8**, we can see that axial solid velocities are similar for all values of e at bed height of 0.27 m but obviously different from each other at bed height of 0.43 m. It indicates that the behaviour of fluid is very complicated and difficult to have similar characters at middle of column. Besides, the trend of axial velocity by e = 1 seriously deviates from others in central region at bed height of 0.60 m. When volume fraction reaches its maximum, axial solid velocity reaches its minimum. And it is the coupling effect of upward

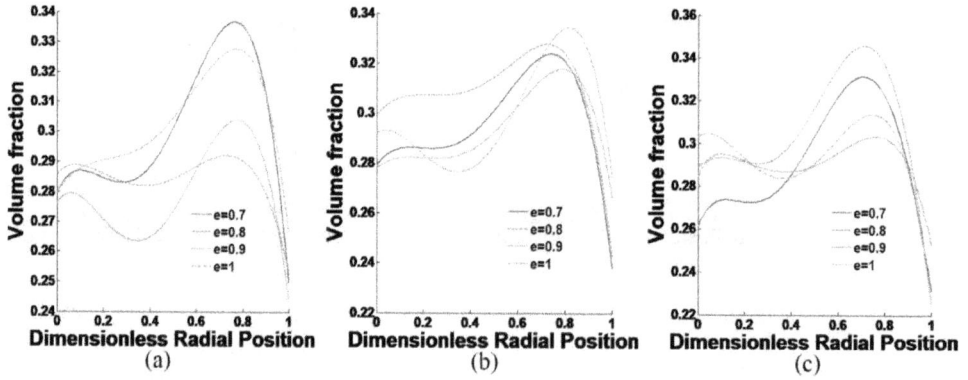

Figure 7. Wall effects on volume fraction at three different bed heights of (a) H = 0.27 m; (b) H = 0.43 m; (c) H = 0.60 m at a superficial liquid velocity of 0.10 m/s.

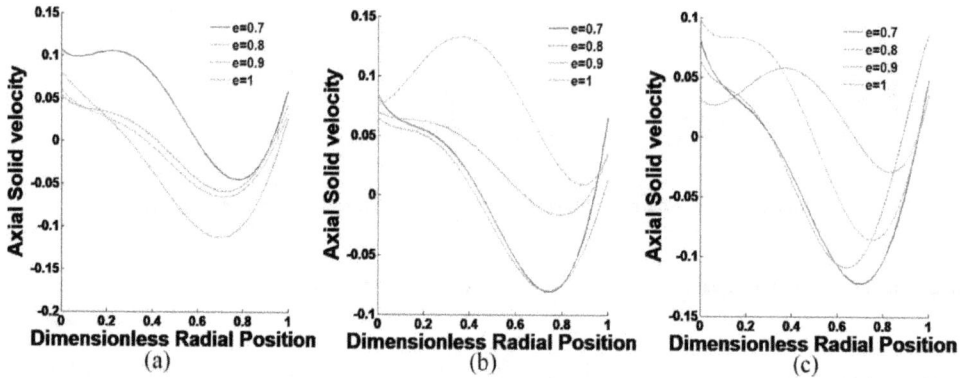

Figure 8. Wall effects on axial solid velocity at three different bed heights of (a) H = 0.27 m; (b) H = 0.43 m; (c) H = 0.60 m at a superficial liquid velocity of 0.10 m/s.

fluid and downward fluid that gathers the particles together near the wall, and the volume fraction reach its maximum.

3.6. Comparison of restitution coefficient

In the particle collision process, particles collide and rebound with energy dissipation during the contact. The ratio between relative velocities after and before collision is the restitution coefficient. In the gas–solid flow system, restitution coefficient is usually a constant, while in the liquid–solid flows, such restitution coefficient is considered varying with the change of Stokes number of particles. In the discrete particle model, the effect of Stokes number on restitution coefficient could be well considered with a varying value; however, in the TFM, we take it as different constant values, so the effect of the relative velocity variation due to the viscosity of fluid could be taken into consideration. In this work, the restitution coefficient varies from 0.7 to 1.0; the effect of different restitution coefficients on particle volume fraction distribution is listed in **Figure 9** for different heights of the bed. Different restitution coefficient can generate different particle volume fraction distribution along the radial direction. Particles with restitution coefficient of 1.0 generate the highest volume fraction in the bed. For other restitution coefficient values, the value of 0.7 generates a comparatively higher particle volume fraction distribution due to energy dissipation during colliding process and makes particle difficult to be transported by the fluid. **Figure 10** shows the effect of different restitution coefficient on axial solid velocity distribution along radial position. The biggest restitution coefficient of 1.0 corresponds to the lowest axial solid velocity distribution at lower part of the bed as **Figure 10(a)** and **(b)** shows. Since particles at the lower part has higher volume fraction, particles tend to have more opportunities to collide, thus along with the velocity distribution, this tends to generate more energy dissipation. At the higher part of the bed, particles near the walls have higher volume fraction and lower velocity distribution due to the frictional resistance of the walls.

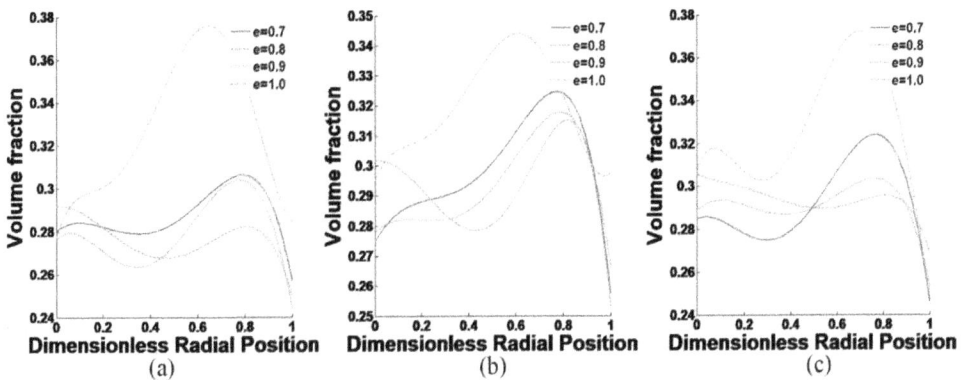

Figure 9. Restitution coefficient on volume fraction at three different bed heights of (a) H = 0.27 m; (b) H = 0.43 m; (c) H = 0.60 m at a superficial liquid velocity of 0.10 m/s.

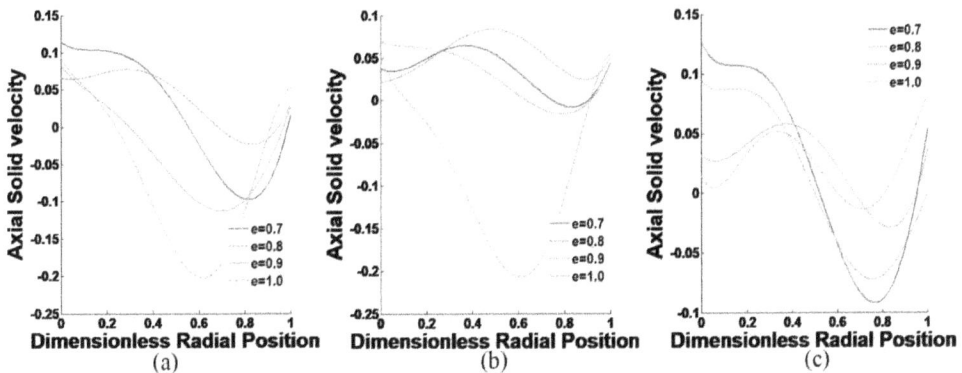

Figure 10. Restitution coefficient of axial solid velocity at three different bed heights of (a) H = 0.27 m; (b) H = 0.43 m; (c) H = 0.60 m at a superficial liquid velocity of 0.10 m/s.

3.7. Comparison of radial distribution

Figure 11 shows the radial distribution of volume fraction at three different bed heights under prediction of TFM-KTGF model with different radial distribution function formations. The Syamlal-O'Brien [10] radial distribution function presents the lowest volume fraction near the walls at three different bed heights, and the Lun [21] radial distribution function model provides the highest volume fraction at most of the radial directions for higher heights in the bed. It is clear that the distribution function can generate obvious different solid volume fraction distribution; however, the distribution tendencies are the same.

Figure 12 shows the radial distribution of axial solid velocity at three different bed heights under prediction of TFM-KTGF model for different radial distribution function formations. The Syamlal-O'Brien radial distribution function presents the lowest axial solid velocity near the walls at three different bed heights. Particle axial velocity distribution shows a decrease of

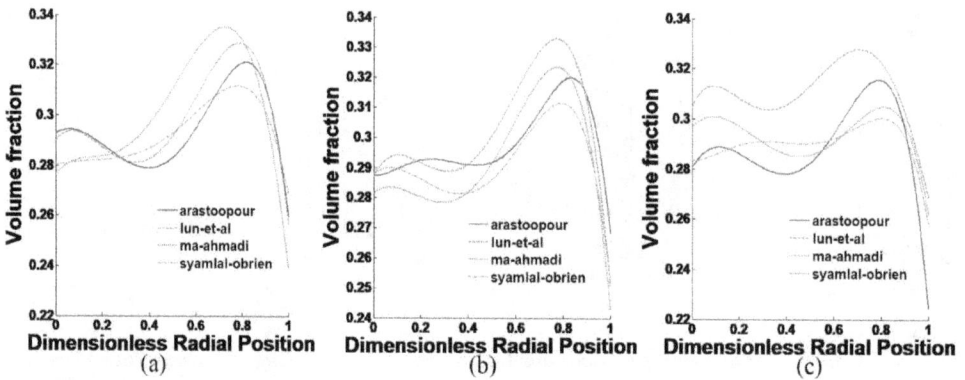

Figure 11. Radial distribution of volume fraction at three different bed heights of (a) H = 0.27 m; (b) H = 0.43 m; (c) H = 0.60 m at a superficial liquid velocity of 0.10 m/s.

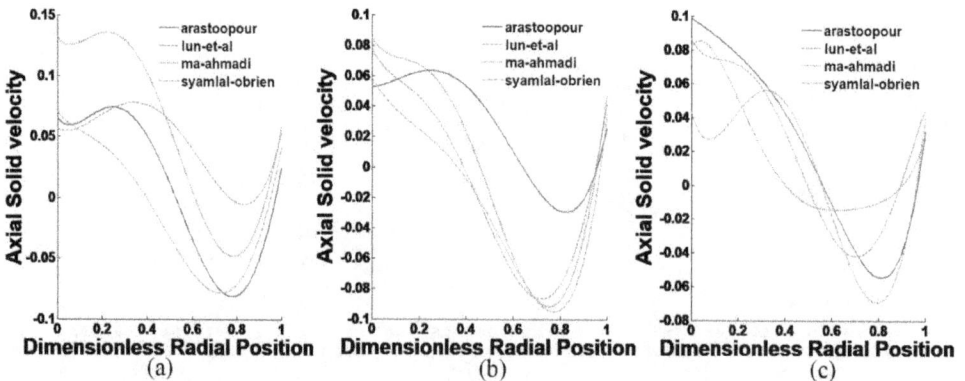

Figure 12. Radial distribution of axial solid velocity at three different bed heights of (a) H = 0.27 m; (b) H = 0.43 m; (c) H = 0.60 m at a superficial liquid velocity of 0.10 m/s.

solid velocity from the centre to the near-wall region and then increase near the walls of the bed. Such distribution pattern agrees well with the previous experimental results.

3.8. Fluctuations at different liquid velocities

Fluctuation of solid phase mean volume fraction at three different liquid velocities of 0.07, 0.10 and 0.13 m/s at the bed height of 0.27 m between 80 and 100 s is shown in **Figure 13**. As one can find from **Figure 13** (a) to (c), the amplitude of the fluctuation increased, and the frequency also increased with the increase of liquid velocity. **Figure 14** shows the fluctuations of mean solid velocity at three different liquid velocities of (a) v = 0.07 m/s; (b) v = 0.10 m/s; (c) v = 0.13 m/s at the bed height of 0.27 m. The trend for the effect of liquid velocity on solid volume fraction fluctuations can also be found in **Figure 14** for the fluctuations of the mean solid velocity.

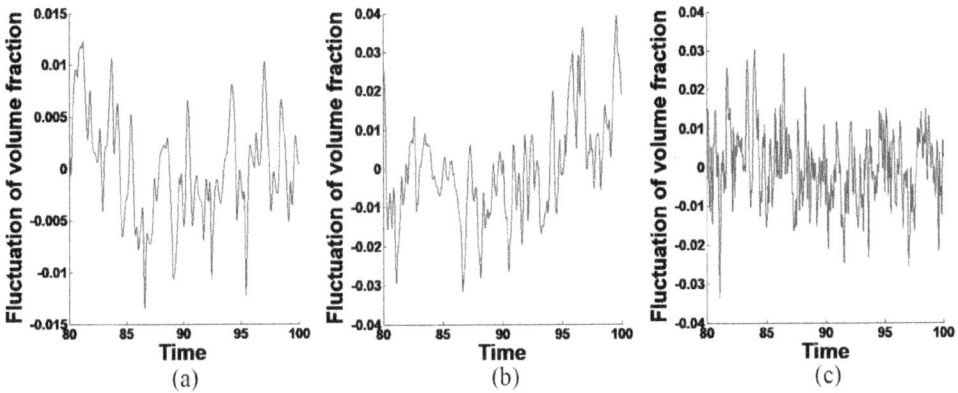

Figure 13. Fluctuation of solid volume fraction at three different liquid velocities of (a) v = 0.07 m/s; (b) v = 0.10 m/s; (c) v = 0.13 m/s at the bed height of 0.27 m.

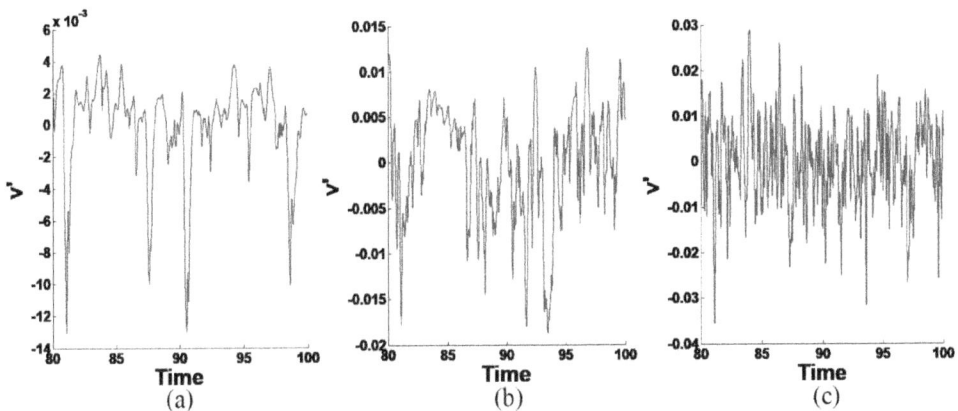

Figure 14. Fluctuation of solid velocity at three different liquid velocities of (a) v = 0.07 m/s; (b) v = 0.10 m/s; (c) v = 0.13 m/s at the bed height of 0.27 m.

3.9. Analysis of granular parameters

Granular pressure distribution as with increasing solid volume fraction for different granular pressure models is shown in **Figure 15(a)**; granular pressure increases with the increase of solid volume fraction to a maximum value for lower volume fractions and then decreases with the increase of solid volume fraction for higher solid volume fractions. Such a distribution is because that the granular pressure is highly related to particle collision, for lower solid volume fraction, increase of solid volume fraction will generate more chance for particles to collide with each other, thus result in a higher granular pressure, while such collision reaches its maximum, the granular pressure will decrease with the increase of solid volume fraction due to collision mechanism is being hindered by more particles per unit volume, and quasi-static contact will play a more important role, thus granular pressure decrease at such solid volume fractions. It is obvious that all the granular pressure models can predict such distribution of granular pressure; however, the quantitative prediction of granular pressure distribution differs for these models. The model proposed by Lun et al. [21] get the highest value while the model of Syamlal and O'Brien [10] get the lowest simulation result. The coefficient of restitution is the ratio of the final to initial velocity difference between two particles after they collide, where one indicates a perfect elastic collision. When it is assumed that the collision is elastic, the granular pressure distribution is totally different from that inelastic collisions where the coefficient of restitution is less than one. As a result, when taking the numerical simulation, it is usually inaccurate to assume that such a liquid–solid system is elastic, since it will result in a falsehood granular pressure distribution.

Granular temperature, $\theta = \frac{1}{3}\left(\langle v_x^2 \rangle + \langle v_y^2 \rangle + \langle v_z^2 \rangle\right)$, is the mean value of the squares of fluctuating velocities at three directions. As indicated previously, with the increase of solid volume fraction, particle collision possibilities increase thus result in a higher granular temperature, while the granular temperature reaches its maximum, more particles per unit volume will

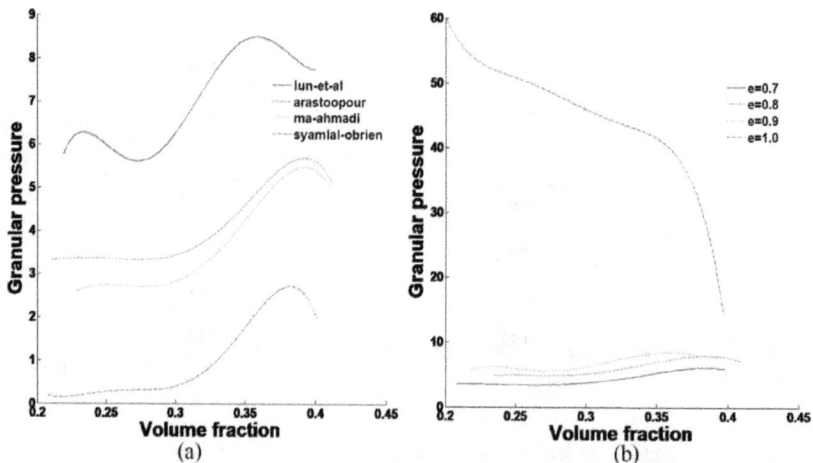

Figure 15. Granular pressure distribution for (a) different granular pressure models, (b) different coefficient of restitution.

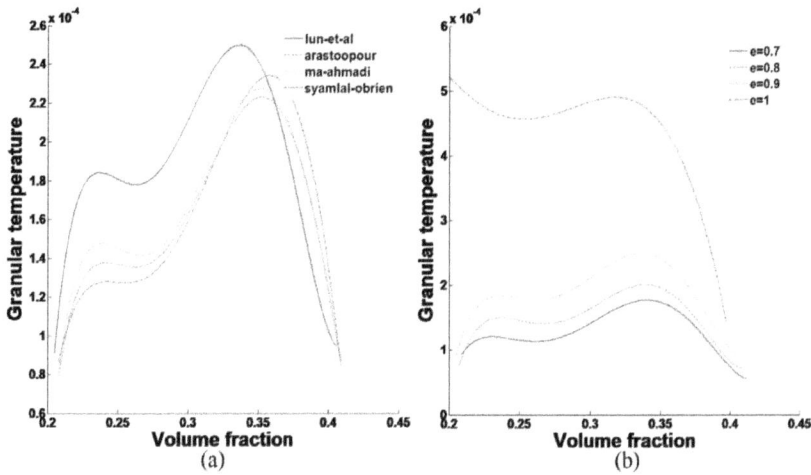

Figure 16. Granular temperature distribution for (a) different granular temperature models, (b) different coefficient of restitution.

bring about a quasi-static contact of particles, thus granular temperature decreases at such solid volume fractions. From **Figure 16(a)**, we can obtain that all of the models for granular pressure predict similar trend of such granular temperature distribution, and the Lun's [21] model provides the highest value for most of the solid volume fraction. When we change the coefficient of restitution coefficient, the elastic collision assumption will result in the highest granular temperature and the trend for such distribution is also unreasonable, since particle collisions in liquid solid flow systems are inelastic, more dissipation will be generated due to fluid drag, fluctuation, etc. As a result, it should be sensitive and careful to select the coefficient of restitution according to the flow system.

4. Conclusions

The TFM combined with the kinetic theory of granular flow (KTGF) is employed to investigate the hydrodynamics of particles in gas–solid as well as liquid–solid fluidized bed. A variety of models and parameters including drag models, granular pressure models, coefficient of restitution are selected when carrying out such numerical simulation. The effect of such models along with the selection of the values for such parameters is comprehensively studied in this work. Numerical investigation of the particle concentration distribution in a liquid–solid fluidized bed is carried out to study the effects of drag force models as well as virtual mass force and lift force in predicting of particle flow characteristics. Different density ratios of solid/liquid, liquid viscosity as well as superficial velocities of liquid phase are also studied to investigate the effects of operating conditions on the distribution of particle concentration.

The predicted axial particle concentration shows a nearly uniform distribution throughout the bed for the investigated particles. Different drag models exhibit various particle velocity distributions indicating that selection of the drag models should be careful. Virtual mass force and lift force should be considered due to the low solid–fluid density ratio. Distribution of granular pressure and granular temperature indicates that elastic assumption for liquid–solid fluidized bed in improper and more energy dissipation due to fluid interstitial effect should be taken into account.

Nomenclature

ε	concentration of each phases (−)
ρ	density (kg/m^3)
p	thermodynamic pressure (N)
$\overline{\tau}_l$	viscous stress tensor of liquid phase (Pa)
μ_t	turbulent viscosity for the liquid phase (N·s/m^2)
k	turbulent kinetic energy (m^2/s^2)
ε	dissipation rate of turbulent kinetic energy (m^2/s^2)
μ_s	shear solid viscosity (Pa·s)
C	particle fluctuating velocity (m/s)
e	coefficient of restitution for particle-particle collisions (−)
L	characteristic length scale (m)
$\varepsilon_{s,\,max}$	maximum particle concentration at random packing (−)
$\mu_{s,\,kin}$	the viscosity induced from particle fluctuations(N·s/m^2)
ϕ	internal friction angle (°)
k_s	conductivity of granular energy
D_{ls}	the rate of energy dissipation per unit volume
$\boldsymbol{u}_s$	velocity vector of solid (m/s)
$\boldsymbol{F}_{vm}$	virtual mass force (N)
$u_{t,w}$	tangential velocity at the wall (N·s/m^2)
$\boldsymbol{u}$	velocity vector (m/s)
$\boldsymbol{g}$	gravity acceleration (N/kg)
β	inter-phase drag coefficient (−)
μ_f	viscosity of liquid phase (N·s/m^2)
μ_l	laminar viscosity for the liquid phase (N·s/m^2)
$\overline{\tau}s$	stress tensor of solid phase (Pa)

ξ_s	bulk solid viscosity (Pa·s)
θ	granular temperature (k)
p_s	solid pressure (Pa)
λ	mean free path of particles (m)
g_0	radial distribution function at contact $(-)$
$\mu_{s,col}$	the viscosity induced from the particle collisions (N·s/m^2)
$\mu_{s,fr}$	the viscosity induced from particle frictions (N·s/m^2)
I_{2D}	the second invariant of deviatoric stress tensor (Pa)
γ_s	rate of dissipation of fluctuation kinetic energy due to particle collisions
u_l	velocity vector of liquid(m/s)
φ	switch function $(-)$
F_{lf}	lift force (N)
θ_w	granular temperature of solid particles at the wall (k)

Author details

Guodong Liu

Address all correspondence to: gdliu@hit.edu.cn

School of Energy Science and Engineering, Harbin Institute of Technology, China

References

[1] Gidaspow D. Multiphase flow and fluidization: Continuum and kinetic theory descriptions. San Diego: Academic Press; 1994

[2] Ding J, Gidaspow D. A bubbling fluidization model using kinetic-theory of granular flow. AICHE Journal. 1990;**36**(4):523-538. DOI: 10.1002/aic.690360404

[3] Benyahia S, Arastoopour H, Knowlton TM, Massah H. Simulation of particles and gas flow behavior in the riser section of a circulating fluidized bed using the kinetic theory approach for the particulate phase. Powder Technology. 2000;**112**(1-2):24-33. DOI: 10.1016/S0032-5910(99)00302-2

[4] van Wachem BGM, Schouten JC, van den Bleek CM, Krishna R, Sinclair JL. Comparative analysis of CFD models of dense gas-solid systems. AICHE Journal. 2001;**47**(5):1035-1051. DOI: 10.1002/aic.690470510

[5] Agrawal K, Loezos PN, Syamlal M, Sundaresan S. The role of meso-scale structures in rapid gas-solid flows. Journal of Fluid Mechanics. 2001;**445**:151-185

[6] Goldschmidt MJV, Kuipers JAM, van Swaaij WPM. Hydrodynamic modelling of dense gas-fluidised beds using the kinetic theory of granular flow: effect of coefficient of restitution on bed dynamics. Chemical Engineering Science. 2001;**56**(2):571-578. DOI: 10.1016/S0009-2509(00)00262-1

[7] Lu H, Gidaspow D, Bouillard J, Liu W. Hydrodynamic simulation of gas–solid flow in a riser using kinetic theory of granular flow. Chemical Engineering Journal. 2003;**95**(1):1-13

[8] Clift R, Grace JR, Weber ME. Bubbles, drops, and particles. San Diego: Academic Press; 1978. pp. 263-264

[9] Wen CY. Mechanics of fluidization. The Chemical Engineering Progress Symposium series. 1966;**62**:100-111

[10] Syamlal M, O'Brien TJ. The derivation of a drag coefficient formula from velocity-voidage correlations. Unpublished Report. 1987

[11] Roy S, Dudukovic MP. Flow mapping and modeling of liquid-solid risers. Industrial & Engineering Chemistry Research. 2001;**40**(23):5440-5454. DOI: 10.1021/ie010181t

[12] Razzak SA, Agarwal K, Zhu JX, Zhang C. Numerical investigation on the hydrodynamics of an LSCFB riser. Powder Technology. 2008;**188**(1). DOI: 42-51. DOI: 10.1016/j.powtec.2008.03.016

[13] Cheng Y, Zhu J. Hydrodynamics and scale-up of liquid-solid circulating fluidized beds: Similitude method vs CFD. Chemical Engineer Science. 2008;**63**(12):3201-3211. DOI: 10.1016/j.ces.2008.03.036

[14] Cheng Y, Zhu JX. CFD modelling and simulation of hydrodynamics in liquid-solid circulating fluidized beds. Canadian Journal Chemical Engineering. 2005;**83**(2):177-185

[15] Liu GD, Wang P, Wang S, Sun LY, Yang YC, Xu PF. Numerical simulation of flow behavior of liquid and particles in liquid-solid risers with multi scale interfacial drag method. Advanced Powder Technology. 2013;**24**(2):537-548. DOI: 10.1016/j.apt.2012.10.007

[16] Liu GD, Wang P, Yu F, Zhang YN, Guo WT, Lu HL. Cluster structure-dependent drag model for liquid-solid circulating fluidized bed. Advanced Powder Technology. 2015;**26**(1):14-23. DOI: 10.1016/j.apt.2014.07.018

[17] Liu GD, Yu F, Lu HL, Wang S, Liao PW, Hao ZH. CFD-DEM simulation of liquid-solid fluidized bed with dynamic restitution coefficient. Powder Technology. 2016;**304**:186-197. DOI: 10.1016/j.powtec.2016.08.058

[18] Xie L, Luo ZH. Modeling and simulation of the influences of particle-particle interactions on dense solid-liquid suspensions in stirred vessels. Chemical Engineering Science. 2018;**176**:439-453. DOI: 10.1016/j.ces.2017.11.017

[19] Rahaman MS, Choudhury MR, Ramamurthy AS, Mavinic DS, Ellis N, Taghipour F. CFD modeling of liquid-solid fluidized beds of polydisperse struvite crystals. International Journal of Multiphase Flow. 2018;**99**:48-61. DOI: 10.1016/j.ijmultiphaseflow.2017.09.011

[20] Ozel A, de Motta JCB, Abbas M, Fede P, Masbernat O, Vincent S, et al. Particle resolved direct numerical simulation of a liquid-solid fluidized bed: Comparison with experimental data. International Journal of Multiphase Flow. 2017;**89**:228-240. DOI: 10.1016/j.ijmulti phaseflow.2016.10.013

[21] Lun CKK, Savage SB. A Simple Kinetic Theory for Granular Flow of Rough, Inelastic, Spherical Particles. Journal of Applied Mechanics. 1987;**54**(1):47-53

[22] Bagnold RA. Experiments on a gravity-free dispersion of large solid spheres in a Newtonian fluid under shear. Proceedings of the Royal Society of London. 1954;**225**(1160):49-63

[23] Iddir H, Arastoopour H. Modeling of multitype particle flow using the kinetic theory approach. AICHE Journal. 2005;**51**(6):1620-1632

[24] Ahmadi G, Ma D. A Thermodynamical formulation for dispersed multiphase turbulent flows-1: Basic theory. International Journal of Multiphase Flow. 1990;**16**(2):323-340. DOI: 10.1016/0301-9322(90)90062-N

[25] Ma D, Ahmadi G. A Thermodynamical formulation for dispersed multiphase turbulent flows-2: Simple shear flows for dense mixtures. International Journal of Multiphase Flow. 1990;**16**(2):341-351. DOI: 10.1016/0301-9322(90)90063-O

[26] Lebowitz JL. Exact solution of generalized Percus-Yevick equation for a mixture of hard spheres. Physical Review A: General Physics. 2008;**133**(133):A895

[27] Koch DL, Pope SB. Coagulation-induced particle-concentration fluctuations in homogeneous, isotropic turbulence. Physics of Fluids. 2002;**14**(7):2447-2455. DOI: 10.1063/1.1478562

[28] Ergun S. Fluid flow through packed columns. Chemical Engineering Progress. 1952;**48**(2): 89-94

[29] van Wachem BGM, Almstedt AE. Methods for multiphase computational fluid dynamics. Chemical Engineering Journal. 2003;**96**(1-3):81-98. DOI: 10.1016/j.cej.2003.08.025

[30] Garside J, Aldibouni MR. Velocity-voidage relationships for fluidization and sedimentation in solid-liquid Systems. Journal of Inorganic and Nuclear Chemistry. 1977;**30**(1):133-145

[31] Ishii M, Mishima K. Two-fluid model and hydrodynamic constitutive relations. Nuclear Engineering and Design. 1984;**82**(2):107-126

[32] Drew D, Lahey R. Particulate two-phase flow. Boston: Butterworth-Heinemann; 1993. pp. 509-566

[33] Johnson PC, Jackson R. Frictional-collisional constitutive relations for granular materials, with application to plane shearing. Journal of Fluid Mechanics. 1987;**176**:67-93. DOI: 10.10 17/S0022112087000570

[34] Gibilaro LG, Difelice R, Waldram SP, Foscolo PU. Generalized friction factor and drag coefficient correlations for fluid particle interactions. Chemical Engineering Science. 1985; **40**(10):1817-1823. DOI: 10.1016/0009-2509(85)80116-0

[35] Limtrakul S, Chen J, Ramachandran PA, Duduković MP. Solids motion and holdup profiles in liquid fluidized beds. Chemical Engineering Science. 2005;**60**(7):1889-1900

Probabilistic Settlement Analysis of Granular Soft Soil Foundation in Southern China Considering Spatial Variability

Lin-Chong Huang, Shuai Huang and Yu Liang

Additional information is available at the end of the chapter

http://dx.doi.org/10.5772/intechopen.79193

Abstract

In this chapter, the method of combining the theory of random field and numerical analysis was used to systematically analyze the settlement probability of the soft soil foundation in the south of China, considering the effect of spatial variability of soil parameters. Based on the midpoint discretization and Cholesky decomposition, the cross-correlated non-Gaussian random field of cohesion and internal friction angle was constructed, which had considered the cross-correlation, and a single parameter random field of modulus was also constructed. The Monte-Carlo stochastic finite element program for two-dimensional foundation probabilistic settlement was developed in APDL language. The influence of spatial variability of soil parameters on probability foundation settlement was studied. The results indicate that the foundation settlement increases with the increase of coefficient variation and correlation distance. Modulus is the most important parameter for foundation settlement. The settlement of foundation is more sensitive to the correlation distance in vertical direction. Based on exponential square autocorrelation function, the continuity of random fields is obviously better, and the foundation settlement is larger. On the contrary, the fluctuation of random fields is larger, and the foundation settlement is smaller with single exponential autocorrelation function.

Keywords: foundation settlement, soil spatial variability, random field, autocorrelation function, midpoint discretization

1. Introduction

The soft soil is widely distributed in the coastal areas of southern China, which exhibits high compressibility and low shear strength [1]. With the acceleration of infrastructure construction in the region, many structures are built on soft soil foundation. Therefore, it is of great significance to study the settlement prediction of soft soil foundation. At present, the prediction methods of foundation settlement are mainly classical formula [2, 3] and numerical analysis [4, 5]. However, these two traditional methods have neglected the spatial variability of soil parameters as a result of mineralogical composition, stress history, and deposition process [6]. At present, many scholars have considered the spatial variability of soil parameters when studying on geotechnical engineering. Yan et al. [7] used the field data of Tianjin Port to establish the random field model of the foundation soil, analyzed and obtained the general law of determining the reduction function with the completely unrelated distance method. Li et al. [8] proposed a noninvasive stochastic finite element method for the reliability analysis of underground caverns; the accuracy and efficiency of calculation were improved. Jiang et al. [9] used random field model to characterize the spatial variability of soil hydraulic conductivity, effective cohesion, and internal friction angle. The effects of rainfall intensity, variability of soil parameters, and cross-correlation between parameters on slope reliability were studied. Kenarsari and Chenari [10] simulated soil mass as an anisotropic random field, combined with FLAC2D finite difference model to study the influence of soil spatial variability on settlement of shallow ground. Lo and Leung [11] used Latin hypercube sampling with dependence to simulate the random field, which was coupled with polynomial chaos expansion to approximate the probability density function of model response, and applied it to the reliability analysis of strip foundation and slope. Johari [12] presented a reliability-based analysis of strip-footing settlement by stochastic finite-element method and combined with random finite-element method to improve computational efficiency.

The above researches are to introduce random field theory into geotechnical engineering, considering the spatial variability of soil. There is spatial autocorrelation of soil between any two points in space, which is usually characterized by correlation distance. And the correlation is inversely proportional to the distance between two points. Autocorrelation functions are generally used to solve the correlation distance. Common autocorrelation functions include single exponential (SNX), exponential square (SQX), cosine exponential (CSX), second-order Markov (SMK), and binary noise (BIN) [13, 14]. Unfortunately, many random field researches in geotechnical engineering were assumed to the autocorrelation function of random field simulation. In order to simplify the calculation, the single exponential autocorrelation function was used to characterize the spatial correlation of the soil parameters. There are few studies considering the influence of the selection of autocorrelation function on foundation settlement. In this chapter, the cross-correlated non-Gaussian random field of South China soft soil was simulated by the Cholesky decomposition technique with midpoint discretization, and then a Monte-Carlo stochastic finite element program for probability settlement of two-dimensional foundation was developed, to study the quantitative evaluation of different autocorrelation functions. This chapter mainly studied the influence of the type of autocorrelation function on foundation settlement when considering the variation of parameter variability, correlation distance, and cross-correlation of parameters.

2. Theoretical basis of random field

The spatial variability of soil parameters reflects the unity of correlation and randomness. This characteristic of soil can be well described with the theory of the random field.

2.1. Numerical characteristics

A random field $S(u)$ can be defined as a curve in vectoral space, which is a collection of random variables indexed by a continuous parameter. For the random field, the most important three numerical characteristics are mean (μ), variance (σ^2), and correlation distance (δ) [15].

The variability of parameters and spatial correlation of soil are all the basic properties of geomaterials. Parameter variability is generally described with coefficient of variation, and the correlation can be described by the correlation distance which is expressed as Eq. (1). Its physical meaning is to measure the size of closely related element in the soil. Within the correlation distance, the soil property of two points is completely correlated, and the geotechnical properties of two points are independent outside the related distance. For the homogeneous random field, the mean and variance are constant, and correlation distance depends only on the distance between two points in the space [13].

$$\delta = \lim_{u \to \infty} u\Gamma^2(u) \tag{1}$$

where $\Gamma^2()$ is the variance reduction function, which represents the ratio of the mean variance in the range u space to the point variance of the random field.

2.2. Autocorrelation function

Based on a large number of measured data, the autocorrelation of soil random fields can be directly derived with the sample autocorrelation function, which is expressed as Eq. (2) [16].

$$\rho_{S(\Delta u)} = \rho[S(u), S(u + (\Delta u))] = \frac{COV[S(u), S(u + \Delta u)]}{\sqrt{var[S(u)]}\sqrt{var[S(u + \Delta u)]}} \tag{2}$$

The limited number of field measured data is usually difficult to directly characterize the spatial correlation of soil parameters. Therefore, the theoretical autocorrelation function is used to fit the sample autocorrelation function. Common autocorrelation functions include single exponential (SNX), exponential square (SQX), cosine exponential (CSX), second-order Markov (SMK), and binary noise (BIN). Such five kinds of two-dimensional autocorrelation function expressions and function images are shown in **Table 1**. The difference of these autocorrelation functions is small when the distance between any two points in the space is large. SQX and SMK are isotropic, and their surfaces are smooth. The edges and corners of SNX, CSX, and BIN are clear, and the continuity is poor.

Types	Functional expressions $\rho(\tau_x, \tau_y)$	Function graph ($\delta_x = \delta_y = 1$)
SNX	$\rho(\tau_x, \tau_y) = \exp\left[-2\left(\frac{\tau_x}{\delta_x} + \frac{\tau_y}{\delta_y}\right)\right]$	
SQX	$\rho(\tau_x, \tau_y) = \exp\left[-\pi\left(\frac{\tau_x^2}{\delta_x^2} + \frac{\tau_y^2}{\delta_y^2}\right)\right]$	
CSX	$\rho(\tau_x, \tau_y) = \exp\left[-\left(\frac{\tau_x}{\delta_x} + \frac{\tau_y}{\delta_y}\right)\right]\cos\left(\frac{\tau_x}{\delta_x}\right)\cos\left(\frac{\tau_y}{\delta_y}\right)$	
SMK	$\rho(\tau_x, \tau_y) = \exp\left[-4\left(\frac{\tau_x}{\delta_x} + \frac{\tau_y}{\delta_y}\right)\right]\left(1 + \frac{4\tau_x}{\delta_x}\right)\left(1 + \frac{4\tau_y}{\delta_y}\right)$	
BIN	$\rho(\tau_x, \tau_y) = \begin{cases} \left(1 - \frac{\tau_x}{\delta_x}\right)\left(1 - \frac{\tau_y}{\delta_y}\right) & \begin{array}{l}\tau_x \le \delta_x \\ \tau_y \le \delta_y\end{array} \\ 0 & \text{else} \end{cases}$	

τ_x, τ_y, respectively, represent the relative distance between horizontal and vertical directions of any two points. δ_x, δ_y, respectively, represent the correlation distance between the horizontal and vertical directions.

Table 1. Common analytical models for autocorrelation functions.

3. Random field simulation of soft soil in South China

In practical engineering, the soil generally obeys non-Gaussian distribution, and there is some cross-correlation in the soil parameters. For example, there is a significant negative correlation between soil cohesion and internal friction angle. In this chapter, the cross-correlated non-Gaussian random fields of soft ground in South China were simulated, based on Cholesky decomposition technique with midpoint discretization [17–20].

3.1. Simulation process

The variability of Poisson's ratio and density of soft soil is relatively small. Therefore, the spatial variability of modulus, cohesion, and internal friction angle is only considered in this chapter. The random field considering the cross-correlation between cohesive and internal friction angle is introduced below. Cross-correlated non-Gaussian distribution of random field simulation needs to generate the cross-correlated standard Gaussian random field. The cross-correlated logarithmic random field can be expressed as Eq. (3) [19].

$$S_i(x,y) = \exp\left(\mu_{\ln i} + \sigma_{\ln i} \cdot S_i^D(x,y)\right) \ (i = c, \varphi) \tag{3}$$

where (x, y) represents the position coordinate of the random field space point; $\mu_{\ln i}$, $\sigma_{\ln i}$ represent the mean and variance of the normal variable $\ln i$, respectively, which is solved by Eq. (4); $S_i^D(x,y)$ represents the relevant standard Gaussian random field.

$$\left.\begin{array}{l} \sigma_{\ln i} = \sqrt{\ln\left(1 + (\sigma_i/\mu_i)^2\right)} \\ \mu_{\ln i} = \ln \mu_i - \dfrac{1}{2}\sigma_{\ln i}^2 \end{array}\right\} \tag{4}$$

The cross-correlated non-Gaussian random field simulation focuses on the generation of Gaussian distribution of the relevant standard Gaussian distribution field, $S_i^D(x,y)$. The process is as follows:

(1) The autocorrelation between any two points of the soil is considered, which is characterized by the autocorrelation coefficient matrix K of the soil. K is solved by the theoretical autocorrelation function. The Cholesky decomposition of the autocorrelation coefficient matrix K is performed, $K = L_1 L_1^T$, and the lower triangular matrix L_1 is obtained.

$$K_i = \begin{bmatrix} 1 & \rho_{12}^i & \cdots & \rho_{1n_e}^i \\ \rho_{12}^i & 1 & \cdots & \rho_{2n_e}^i \\ \vdots & \vdots & \ddots & \vdots \\ \rho_{1n_e}^i & \rho_{2n_e}^i & \cdots & 1 \end{bmatrix} \ (i = c, \varphi) \tag{5}$$

where n_e represents the number of random field elements.

(2) Considering the cross-correlation between cohesion and internal friction angle, the cross-correlation coefficient matrix R is used to represent it. Cholesky decomposition of the cross-correlation matrix, $R = L_2 L_2^T$, leads to the lower triangular matrix L_2. Due to the transformation in the random field simulation, theoretically, the correction of R and K needs to be modified according to the Nataf model. However, the difference of the correlation coefficient matrix between Gaussian and lognormal random fields is very small [18]. Take the correction coefficient of 1.

$$R = \begin{bmatrix} 1 & \rho_{c,\varphi} \\ \rho_{c,\varphi} & 1 \end{bmatrix} \tag{6}$$

(3) A set of related standard normal random sample matrices α was derived using Latin hypercube sampling, $\alpha_i = \{\alpha_i^1, \alpha_i^2, \cdots, \alpha_i^{n_e}\}$, $(i = c, \varphi)$. According to Eq. (7), the cross-correlated standard Gaussian random field $S_i^D(x, y)$ is obtained.

$$S_i^D(x, y) = L_1 \cdot \alpha \cdot L_2^T \tag{7}$$

The cross-correlated non-Gaussian random field simulation is completed with the cohesion and friction angle, by taking Eq. (7) into the Eq. (3). The simulation of modulus random field is consistent with the above process, which will not be repeated here. Because it is a single parameter random field, the cross-correlation coefficient need not be considered in the calculation process, and the simulation process is simpler.

3.2. Typical realizations of random fields

Based on MATLAB software, the random field procedure was written according to the process above. A typical South China homogeneous soft soil foundation was adopted to simulate. The size and soil parameter of this foundation were introduced in the Section 4.1. The coefficient of variation of modulus, cohesion, and internal friction angle are 0.3. The cross-correlation coefficient of cohesion and internal friction angle is −0.5. The size of random field elements is 0.5 m, the correlation distance in horizontal, and vertical directions are 40 m and 3 m, respectively. **Figure 1** shows typical realizations of random field of c and φ for five autocorrelation functions.

Figure 1(a), **(b)**, **(c)**, **(e)** and **(f)** shows the typical realizations of random fields of cohesion with five autocorrelation functions, respectively. In these figures, the red regions denote a larger strength parameter value, while the blue regions indicate a smaller strength parameter value. The continuity of random fields based on SQX and SMK is obviously better than the other three kinds of autocorrelation functions. And the fluctuation of the SNX is the largest. This conclusion is consistent with the continuity of the theoretical autocorrelation function in **Table 1**. For **Figure 1(c)** and **(d)**, the distribution of random fields of c and φ is approximately the opposite, where the value of cohesive is large, and the value of internal friction angle is small. The overall trend is negative correlated. The difference between the random fields

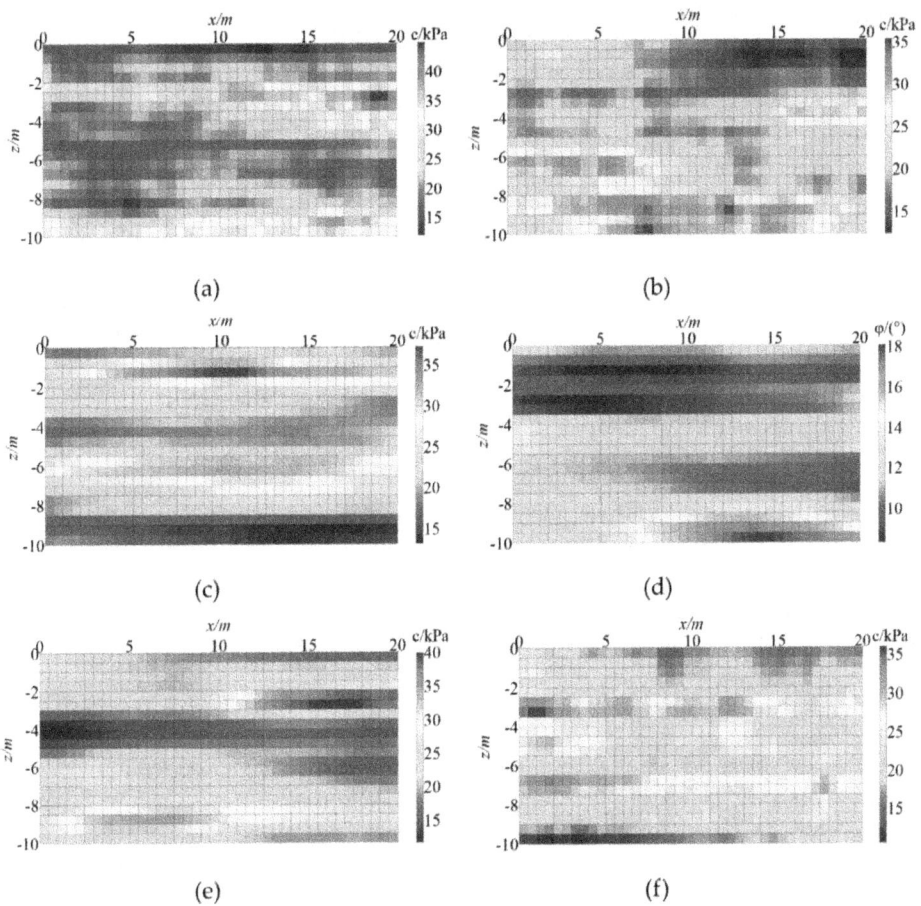

Figure 1. Typical realizations of random fields of c and φ for five autocorrelation functions. (a) SNX, c; (b) CSX, c; (c) SQX, c; (d) SQX, φ; (e) SMK, c; and (f) BIN, c.

established by the five autocorrelation functions is larger. Therefore, it is important to study the influence of autocorrelation function selection on foundation settlement [21, 22].

4. Example of foundation settlement analysis

In this chapter, a typical southern soft soil ground in China was selected. First, a deterministic model was established (mean value of soil parameters), and then, the probabilistic analysis of ground settlement with the random field finite element model of the soil parameters was carried on. The influence of spatial variability of soil parameters and selection of autocorrelation function on foundation settlement was studied.

4.1. Deterministic analysis

Deterministic calculation does not consider the spatial variability of the parameters, which assigns the same soil parameters to each element. Based on ANSYS software, a two-dimensional foundation plane strain model was established. The horizontal width of this model is 20 m, and the vertical depth is 10 m. There is a rigid strip foundation above the foundation soil with a foundation width of 2 m. Foundation geometry and finite element mesh division are shown in **Figure 2**. To facilitate the randomness analysis, the mesh size is consistent with the size of the random field in Section 3.2 (0.5 m), which consisted of 800 elements and 861 nodes. Drucker-Prager criterion is adopted to represent the stress-strain behavior of the soil. The contact surface and target surface are simulated by CONTA172 and TARGE169, respectively [23]. Both lateral boundaries are rollers, and the base is full fixity. There is a concentrated load P = 100 kN on the foundation. Calculated parameters are as follows: cohesion 20 kPa, internal friction angle 12°, unit weight 18 kN/m^3, modulus of deformation 4 MPa, and Poisson's ratio 0.25.

Figure 3 shows the vertical displacement cloud for deterministic calculation. From **Figure 3**, the maximum settlement is 41.18 mm, which occurs just below the rigid strip foundation. In order to verify the accuracy of the model calculation results, the traditional hierarchical design method was adopted, and the theoretical result is 39.1 mm, which is closed to the simulated one, with the error of 5.3%. It shows that the numerical simulation result is reliable.

4.2. Randomness analysis

The spatial variability of modulus, cohesion, and internal friction angle was mainly considered in this chapter [12]. About 30 calculation conditions were designed as shown in **Table 2**. In each condition, the random fields of E, c, and φ were simulated by five kinds of autocorrelation functions. Based on APDL language, the Monte-Carlo stochastic finite element calculation program for two-dimensional foundation was constructed. Specifically, E, c, and φ were defined as input variables, and the values at each random field were brought into the finite

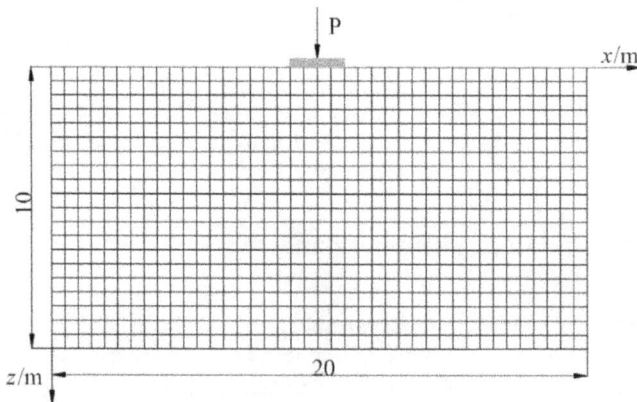

Figure 2. Finite element model of foundation settlement.

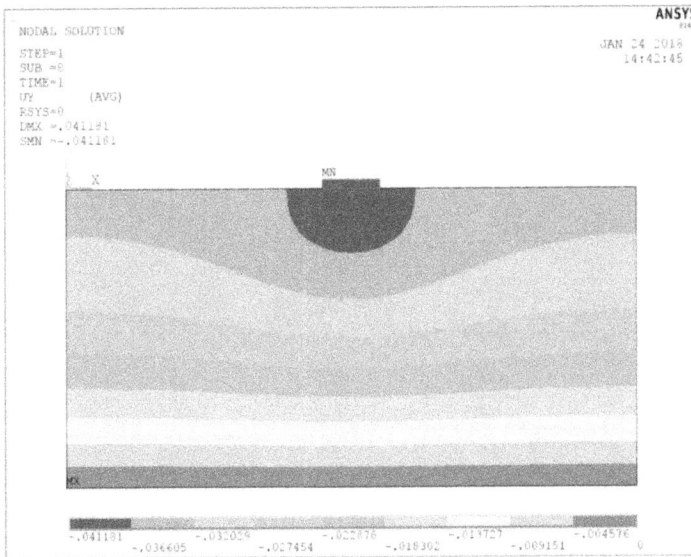

Figure 3. Soft soil vertical displacement cloud image.

element calculation. Then, the results of the finite element calculation were obtained. The maximum vertical displacement (Umax) is the output variable, and the statistics of Umax are required.

Take the RF-E3 condition as an example, where the type of autocorrelation function is SNX. **Figure 6** shows the result of randomness analysis for foundation settlement within the confidence limit of 95%. In **Figure 4(a)**, the mean of maximum settlement of the foundation tends to be stable when the times of simulation reach to 1000. The rest of the calculation conditions also costs the same simulation times. The mean of random analysis in RF-E3 condition is 45.096 mm, which is slightly larger than the result of deterministic analysis. **Figure 4(b)** shows the cumulative distribution curve of the maximum settlement of the foundation. The probability of the maximum settlement of the foundation between the 30 and 60 mm interval is 95%. The foundation settlement can be predicted by probability. If the value of settlement is used as an index of foundation reliability, the failure probability of foundation can be read from the figure.

4.2.1. Analysis of parameter variability

The variability of soil parameters is represented by coefficient of variation (COV) in statistics. The influence of spatial variability on foundation settlement is analyzed by 15 kinds of calculation conditions of RF-E1~RF-φ5. At the same time, the influence of autocorrelation function on foundation settlement is studied.

The effects of coefficient of variation on ground settlement with E, c, and φ are given in **Figure 3**, respectively. It can be seen from the figure that with the increase of coefficient of

Variable	Mean	Coefficient of variation			δ/m		Cross-correlation	Conditions
		E	c	φ	δ_x	δ_y		
E	4 MPa	0.1	0.3	0.3	40	3	−0.5	RF-E1
		0.2						RF-E2
		0.3						RF-E3
		0.4						RF-E4
		0.5						RF-E5
c	20 kPa	0.3	0.1	0.3	40	3	−0.5	RF-c1
			0.2					RF-c2
			0.3					RF-c3
			0.4					RF-c4
			0.5					RF-c5
φ	12°	0.3	0.5	0.1	40	3	−0.5	RF-φ1
				0.2				RF-φ2
				0.3				RF-φ3
				0.4				RF-φ4
				0.5				RF-φ5
δ_x	—	0.3	0.3	0.3	20	3	−0.5	RF-x1
					30			RF-x2
					40			RF-x3
					50			RF-x4
					60			RF-x5
δ_y	—	0.3	0.3	0.3	40	1	−0.5	RF-y1
						2		RF-y2
						3		RF-y3
						4		RF-y4
						5		RF-y5
$\rho_{c,\varphi}$	—	0.3	0.3	0.3	40	3	−0.7	RF-ρ1
							−0.5	RF-ρ2
							0	RF-ρ3
							0.3	RF-ρ4
							0.5	RF-ρ5

Table 2. Calculation conditions.

variation of soil parameters, the mean of the maximum settlement also increases, and all the mean of randomness analysis are larger than the result of deterministic analysis, which indicates that the parameter variability of soil has an important influence on foundation

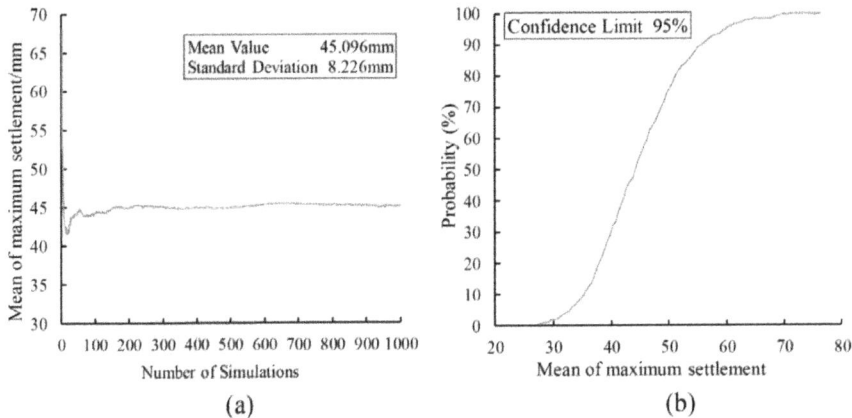

Figure 4. Result of random analysis in RF-E3 condition. (a) Convergence curve and (b) cumulative distribution curve.

settlement. In other words, traditional deterministic analysis underestimates foundation settlement. It is necessary to consider the variation of soil parameters in engineering practice. Contrast the rangeability of the mean of maximum settlement in **Figure 5(a)–(c)**, the curve of modulus changes larger than cohesion and internal friction angle obviously, which means that the parameter sensitivity, $E > \varphi > c$. The influence trend of different autocorrelation function on foundation settlement is basically the same. The mean of maximum settlement obtained by SQX was largest and the SNX was smallest. With the increase of coefficient of variation, the difference of the calculated results with the five autocorrelation functions becomes greater. In **Figure 5(a)**, the difference of settlement calculated by different autocorrelation functions is only 0.1 mm when $COV_E = 0.1$. The difference value increases to 2 mm when $COV_E = 0.5$, which accounts for 20% of the settlement variation value (10 mm) caused by parameter variability. This indicates that the influence of autocorrelation function should be considered when the coefficient of variation becomes larger. As the coefficient of parameter variation increases, the discreteness of random fields increases. These facts indicating the increase in the probability of the appearance of element with low value will cause the increase of foundation settlement. Besides, the smoothness and continuity of the random field by SNX is poor; thus, the elements with low value are discrete. The stability of foundation calculated by SNX is improved, and the foundation settlement calculated by it comes to the smallest.

4.2.2. Analysis of spatial correlation

Correlation distance is one of the important parameters to characterize the spatial variability of soil parameters [24]. The influence of horizontal correlation distance (δ_x) and vertical correlation distance (δ_y) on foundation settlement is studied. About 10 calculation conditions of RF-x1~RF-y5 are set. The random field model is degraded into random variable model when the correlation distance of all directions approaches infinity. Thus, the parameters are completely correlated to the model area.

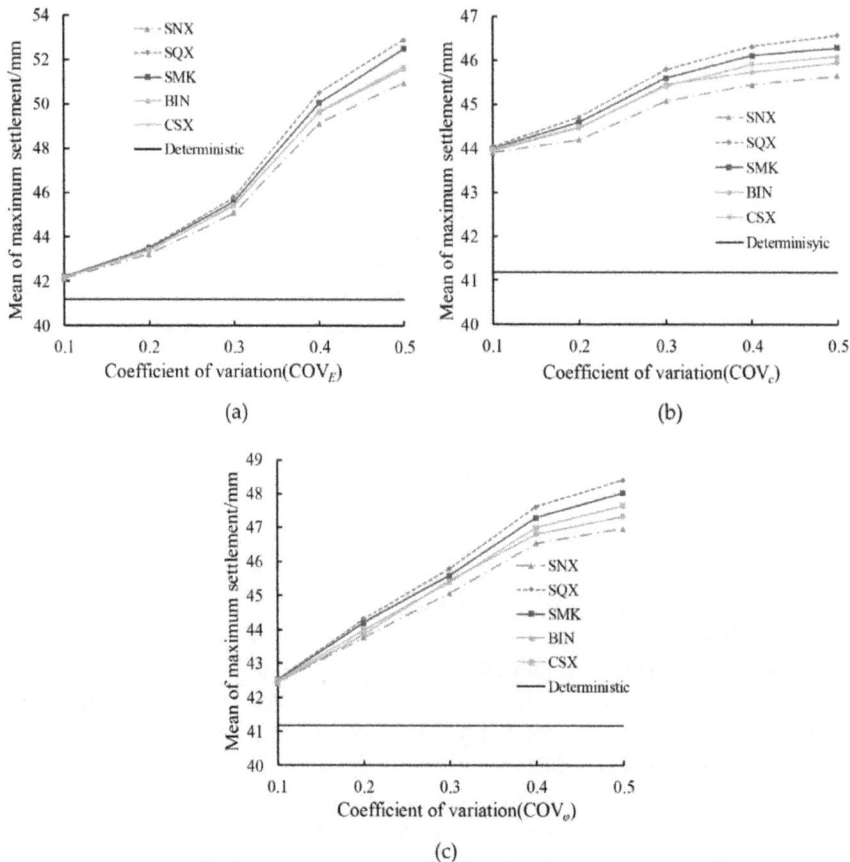

Figure 5. Curve of the mean of maximum settlement with coefficient of variation. (a) Modulus, (b) cohesion, and (c) internal friction angle.

Figure 6 shows the effect of correlation distance on the mean of maximum settlement. The black line in the figure represents the result of the random variable model. The mean of maximum settlement increases with the increase of the correlation distance, which gradually reaches to convergence. The influence of vertical correlation distance on settlement is more significant than that of horizontal correlation distance. It is necessary to simulate the spatial variability of soil parameters with the anisotropic random field. The results of random fields are less than that of the random variable model (46.91 mm). It indicates that ignoring the spatial variability of the soil will lead to the overestimation of the settlement of the foundation. The mean of maximum settlement obtained by SQX was largest and the one obtained by SNX was smallest. As the correlation distance increases, the continuity of the random field will be significantly improved. The elements with low value are also distributed continuously, which is equivalent to the formation of weak intercalated layer in the foundation. The stability of foundation is reduced and the foundation settlement increases. Compared with other

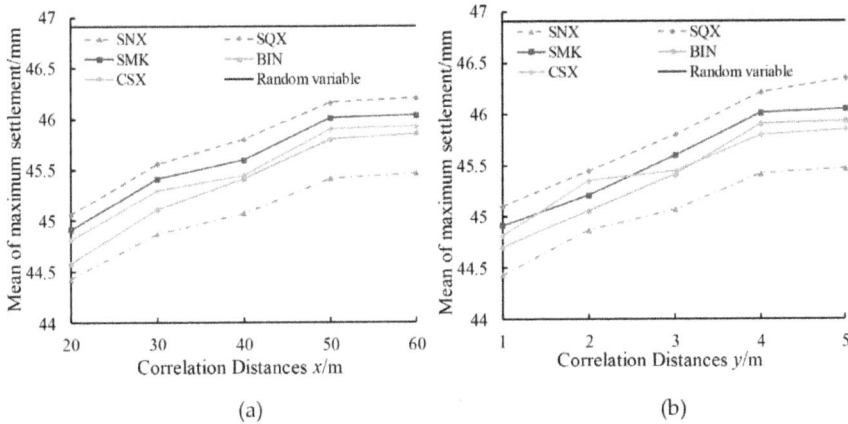

Figure 6. Curve of the mean of maximum settlement with correlation distance. (a) Horizontal correlation distance and (b) vertical correlation distance.

autocorrelation functions, the continuity of the random field by SNX autocorrelation function is poor. Thus, the foundation settlement by SNX comes to the smallest.

In order to incorporate the dependence between the strength parameters, the cross-correlation coefficient ($\rho_{c,\varphi}$) is needed. The study shows that there is a significant negative correlation between c and φ [25]. **Figure 7** shows the effect of cross-correlation between cohesion and friction angle on foundation settlement. With the increase of cross-correlation coefficient, the mean of

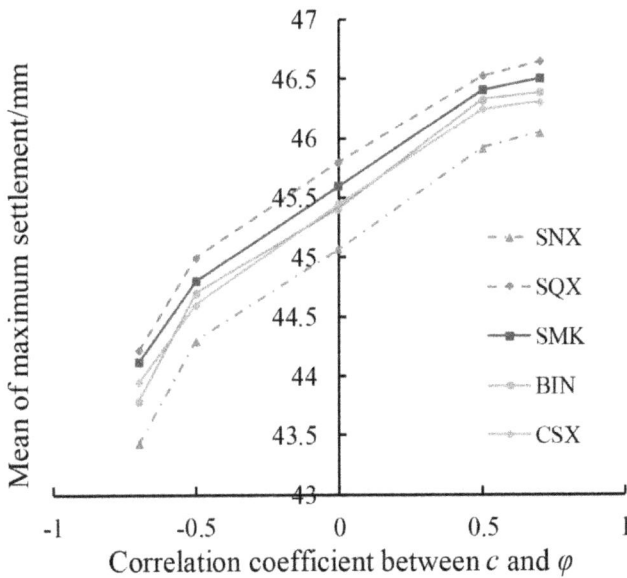

Figure 7. Effect of cross-correlation between cohesion and friction angle on settlement.

maximum settlement increases. This indicates that neglecting the negative correlation between cohesion and internal friction angle will overestimate the settlement of foundation. Considering the negative correlation between cohesion and friction angle, the increase of cohesion corresponds to the decrease of friction angle, which leads to the decrease of the total shear strength variance of soil. The stronger the negative correlation is, the smaller the variance of total shear strength parameters is, which means the small scale of fluctuation of random fields. Thus, the foundation settlement is decreased. The maximum settlement value can be obtained by SQX; the value of SNX is smaller than the other autocorrelation functions obviously.

In summary, the selection of autocorrelation function has obvious influence on the analysis of foundation settlement. The influence trend is basically consistent with the change of statistical parameters of random fields. The settlement value selected by SQX is the largest, and the settlement value selected by SNX is the smallest. In other words, the results of foundation settlement are safer for the designers based on SQX.

5. Conclusions

This chapter combined Cholesky decomposition midpoint method with Monte-Carlo method. The calculation method of two-dimensional ground settlement was obtained based on random field theory. Considering the influence of the autocorrelation function selection in the random field simulation, several conclusions are drawn from this study:

1. Based on the Cholesky decomposition technique with midpoint discretization, the cross-correlated non-Gaussian random fields considering cross-correlation and the independent non-Gaussian random fields are convenient to simulate. The random fields are easier to be introduced into the stochastic finite element model. By changing the type of autocorrelation function in simulation, the influence of the selection of autocorrelation function on foundation settlement is studied. Combined with the typical realization of the random field in Section 3.2, the mechanism of influence on foundation settlement caused by statistics of soil parameters and the type of autocorrelation function can be further explored.

2. The variability of soil parameters has a significant influence on the calculation results of foundation settlement, and the results of randomness analysis are larger than the results of deterministic analysis. The mean value of maximum settlement increases with the variation coefficient of the parameters, and the modulus E of soil affects the calculated value of foundation settlement most. Therefore, the variability of soil parameters should be considered in the calculation of foundation settlement.

3. Spatial correlation of soil has a significant impact on the calculation of foundation settlement. The larger the correlation distance is, the larger the maximum settlement of foundation is. The settlement of foundation is more sensitive to the correlation distance in vertical direction. The mean of maximum settlement increases with the increase of the cross-correlation coefficient between cohesion and internal friction angle.

4. The selection of different autocorrelation functions has a significant effect on foundation settlement; the values of settlement based on SQX and SMK are larger, and that based on SNX and BIN is smaller. The result of SNX is significantly smaller than that of the other types. With the increase of coefficient of variation, the influence of the selection of autocorrelation function on the settlement value also increases.

Acknowledgements

This work was supported by the National Natural Science Foundation of China under Grant 51708564 & Grant 51678578; China Postdoctoral Science Foundation under Grant 2018M633223; the Guangdong Natural Science Foundation of China under Grant 2016A030313233; the Guangzhou Science & Technology Program of China under Grant 201804010107 & Grant 201704020139; and the Department of Communications of Guangdong Province of China under Grant 2016-02-026.

Author details

Lin-Chong Huang, Shuai Huang and Yu Liang*

*Address all correspondence to: liangyu25@mail.sysu.edu.cn

Sun Yat-sen University, Guangzhou, China

References

[1] Huang LC, Zhou CY, Li WH. Modeling the microstructure random fields of soft soil in the south of China. Geotechnical Special Publication. 2014;236:495-501. DOI: 10.1061/9780784 413388.051

[2] Yang GH, Li J, Jia K, et al. Improved settlement calculation method for engineering practice. Chinese Journal of Rock Mechanics and Engineering. 2017;36(S2):4229-4234. DOI: 10.13722/j.cnki.jrme.2016.0681

[3] Yang GH, Yao LN, Jiang Y, et al. Practical method for calculating nonlinear settlement of soft ground based on e-p curve. Chinese Journal of Geotechnical Engineering. 2015;37(2): 242-249. DOI: 10.11779/CJGE201502005

[4] Wang S, Qi J, Yu F, et al. A novel modeling of settlement of foundations in permafrost regions. Geomechanics and Engineering. 2016;10(2):225-245. DOI: 10.12989/gae.2016.10.2.225

[5] Hou JF, Chen J, Kou XQ. Numerical analysis of soft soil ground consolidation settlement. Applied Mechanics and Materials. 2014;638-640:503-506. DOI: 10.4028/www.scientific. net/AMM.638-640.503

[6] Ali A, Huang J, Lyamin AV, et al. Simplified quantitative risk assessment of rainfall-induced landslides modelled by infinite slopes. Engineering Geology. 2014;**179**(10):102-116. DOI: 10.1016/j.enggeo.2014.06.024

[7] Yan SW, Guo LP, Cao YH. Regularity of determination of reduction function of variance in reliability analysis of geotechnical engineering. Rock and Soil Mechanics. 2014;**35**(8):2286-2292. DOI: 10.16285/j.rsm.2014.08.014

[8] Li DQ, Jiang SH, Chen YF, et al. Reliability analysis of serviceability performance for an underground cavern using a non-intrusive stochastic method. Environmental Earth Sciences. 2014;**71**(3):1169-1182. DOI: 10.1007/s12665-013-2521-x

[9] Jiang SH, Li DQ, Zhou CB, et al. Reliability analysis of unsaturated slope considering spatial variability. Rock and Soil Mechanics. 2014;**35**(9):2569-2578. DOI: 10.16285/j.rsm.2014.09.012

[10] Kenarsari AE, Chenari RJ. Probabilistic settlement analysis of shallow foundations on heterogeneous soil stratum with anisotropic correlation structure. In: Proceedings of the IFCEE 2015. International Foundations Congress and Equipment Expo 2015; 17-21 March 2015; San Antonio, TX, USA: IFCEE; 2015. p. 1905-1914

[11] Lo MK, Leung YF. Probabilistic analyses of slopes and footings with spatially variable soils considering cross-correlation and conditioned random field. Journal of Geotechnical & Geoenvironmental Engineering. 2017;**143**(9):1-12. DOI: 10.1061/(ASCE)GT.1943-5606.0001720

[12] Johari A, Sabzi A. Reliability analysis of foundation settlement by stochastic response surface and random finite element method. Scientia Iranica. 2017;**24**(6). DOI: 10.24200/sci.2017.4169

[13] Lin J, Cai GJ, Zou HF, et al. Assessment of spatial variability of Jiangsu marine clay based on random field theory. Yantu Gongcheng Xuebao/Chinese Journal of Geotechnical Engineering. 2015;**37**(7):1278-1287. DOI: 10.11779/CJGE201507014

[14] Cao Z, Wang Y. Bayesian model comparison and selection of spatial correlation functions for soil parameters. Structural Safety. 2014;**49**:10-17. DOI: 10.1016/j.strusafe.2013.06.003

[15] Zhang J, Huang HW, Phoon KK. Application of the kriging-based response surface method to the system reliability of soil slopes. Journal of Geotechnical & Geoenvironmental Engineering. 2013;**139**(4):651-655. DOI: 10.1061/(ASCE)GT.1943-5606.0000801

[16] Wu SH, Ou CY, Ching J, et al. Reliability-based design for basal heave stability of deep excavations in spatially varying soils. Journal of Geotechnical & Geoenvironmental Engineering. 2012;**138**(5):594-603. DOI: 10.1061/(ASCE)GT.1943-5606.0000626

[17] Jiang SH, Dian-Qing LI, Zhou CB, et al. Slope reliability analysis considering effect of autocorrelation functions. Chinese Journal of Geotechnical Engineering. 2014;**36**(3):508-518. DOI: 10.11779/CJGE201403014

[18] Pan Q, Dias D. Probabilistic evaluation of tunnel face stability in spatially random soils using sparse polynomial chaos expansion with global sensitivity analysis. Acta Geotechnica. 2017;**13**:1-15. DOI: 10.1007/s11440-017-0541-5

[19] Zhu H, Zhang LM, Xiao T, et al. Generation of multivariate cross-correlated geotechnical random fields. Computers and Geotechnics. 2017;**86**:95-107. DOI: 10.1016/j.compgeo.2017.01.006

[20] Huang L, Tao C, Yu J, et al. Modelling the microstructure random fields of soft soil under the scale optimized retinex algorithm and microscopic image enhancement. Journal of Intelligent Fuzzy Systems. 2017;**33**(5):1-11. DOI: 10.3233/JIFS-169342

[21] Hekmatzadeh AA, Zarei F, Johari A, et al. Reliability analysis of stability against piping and sliding in diversion dams, considering four cutoff wall configurations. Computers and Geotechnics. 2018;**98**:217-231. DOI: 10.1016/j.compgeo 2018.02.019

[22] Yu L, Chenyuan T, Bingcheng Z, Shuai H, Linchong H. Submicron structure random field on granular soil material with retinex algorithm optimization. In: Proceedings of Powders and Grains 2017-8th International Conference on Micromechanics of Granular Media; 3–7 July 2017; Montpellier. EPJ Web of Conferences 140, 12013; 2017. pp. 1-4

[23] Jayalekshmi BR, Jisha SV, Shivashankar R. Analysis of foundation of tall R/C chimney incorporating flexibility of soil. Journal of the Institution of Engineers. 2017;**98**(1):1-7. DOI: 10.1007/s40030-017-0218-y

[24] Hicks MA, Li Y. Influence of length effect on embankment slope reliability in 3D. International Journal for Numerical and Analytical Methods in Geomechanics. 2018;**42**(1):891-915. DOI: 10.1002/nag.2766

[25] Cai JS, Yan EC, Yeh TCJ, et al. Effect of spatial variability of shear strength on reliability of infinite slopes using analytical approach. Computers and Geotechnics. 2017;**81**:77-86. DOI: 10.1016/j.compgeo.2016.07.012

The Effects of Sediment Size and Concentration on the Rheological Behavior of Debris Flows

Leonardo Schippa

Additional information is available at the end of the chapter

http://dx.doi.org/10.5772/intechopen.79841

Abstract

Sediment concentration, size, and distribution of grains play a relevant role defining the rheology of many geophysical flows. Experiments on slurries consisting of fine-grained and coarse-grained reconstituted debris flow mixtures having bulk volume concentration ranging from 0.32 to 0.42 are examined. The mixtures exhibit a typical yielding non-Newtonian flow behavior. Sediment concentration influences the rheological behavior of the mixtures, leading to dilatant or pseudoplastic flow. A generalized Herschel-Bulkley rheological model well represents the experimental data, whereas power index and consistent coefficient are expressed as a function of sediment concentration (i.e., void ratio). The presence of coarse grain fraction mainly influences yield stress. Increasing the relative content of coarser fraction, with respect to the finer fraction, leads to a diminishing of yield stress. Keeping constant the finer sediment content, the more relevant coarse fraction is the higher yield stress results.

Keywords: debris flows, fine-grained mixtures, coarse-grained mixtures, Herschel-Bulkley model, yield stress

1. Introduction

Granular-fluid mixtures are commonly present in natural flows, such as debris flows or mud flows, whose difference mainly lies in the sediment fine fraction content, which is determinant for the fluid rheological behavior.

Over the last decades, the risk of such geophysical phenomena has increased enormously because of the effects of climate change, and the effort to describe the flow properties has increased too, motivating experimental, numerical, and theoretical studies. Nevertheless, it is still an open question to define the universal features for different flow configuration. Even when using state-of-the-art technologies, it is still difficult to extract common features or a general trend for different flow configurations [1]. The difficulty mainly refers to the uncertainty in defining the most appropriate constitutive equations for the flowing materials, and the knowledge of the rheological behavior of these mixtures is crucial in any run out modeling, to assess the travel distance and the depositional area.

Indeed, debris flows behave as a concentrated grain-fluid mixture of variously assorted particles during the flow [2], and its bulk flow properties can be assessed from the study of the involved soil-liquid mixture accounting for the effects due to particle size distribution and solid volumetric concentration above all [3–8].

In literature, many different models for both dry granular flows and fluid-granular mixtures may be found but they do not provide a unique rheological formula for the mixture [1, 2, 9–13]. One of the most popular approaches considers the debris flow as a non-Newtonian fluid with an empirical Bingham rheology. This methodology shows good results in case of viscous flow (e.g., pure mudflows) [14–16]; conversely, it is not suitable for the case of noncolloidal particles involved in the flowing mixture (i.e., granular flow) [17], in which not only the fluid viscosity but also the grain-fluid interactions have to be taken into account.

Solid volumetric concentration and particle size distribution greatly influence the behavior of granular suspensions: usually, the finest particles are very sensitive to Brownian motion effects or colloidal forces, whereas, coarse particles experience frictional or collisional contacts and hydrodynamic forces. Therefore, the bulk behavior of particle suspensions is very complex and depends on many parameters: solid volumetric concentration, size and shape of the particles, size distribution, the nature of the interstitial fluid, etc.

Accounting for the presence of a large range in sediment size in natural debris flow, the first step may be to understand how the finer (i.e., colloidal fraction) and the coarser (i.e., silty and sandy) fractions contribute to the rheology of the mixture.

Sengun and Probstein [18] carried out experimental investigations and theoretical analysis on coal slurries. They observed that on one hand, the fine (colloidal) fraction seems to perform independently of the coarse fraction and that the fluid matrix, composed by the interstitial liquid and the finest fraction, confers most of its rheological characteristics to the bulk mixture. On the other hand, the coarser particles significantly contribute to the viscosity variation via processes of hydrodynamic dissipation. The experimental work performed by Coussot and Piau [19] on natural debris flows mixtures confirmed that the amount of finest fraction influences the main rheological parameters of the entire suspensions, and that the yield stress strongly varies with the amount of coarse particles. It is in agreement with the observations of Ancey and Jorrot [20] derived from their laboratory experience on coarse particles dispersed in a clay suspension. In fact, they put in evidence that the fine-grained fraction is responsible of the rheological behavior of the bulk mixture, if large particle fraction is smaller than fine particle fraction. Ancey and Jorrot [20] also illustrated that the grain size distribution has relevant effects on the yield stress value: it increases proportionally to the solid concentration of coarse fraction if it is dominant in bulk volume of the mixture.

The works carried out in the last decades on a large amount of collected data indicate the Hershel-Bulkley model as the most appropriate to describe the rheological behavior of simple yield stress fluid in a large range of shear rate (i.e., 10^{-1}–10^2 s^{-1}) [21]. Referring to the steady state flow-like regime of debris flows, rheological parameters may be expressed as a function of the bulk volume sediment concentration, whereas the yield stress greatly depends not only on the sediment concentration but also on the relative content of finer and coarser grain [22].

This chapter presents recent experiments [22] on reconstituted debris flows mixture, stressing the effects on the rheological behavior due to the sediment concentration and the presence of coarse-grained fraction.

Experimental activities carried out with rotational rheometer and inclined plane are presented separately. The former was mainly oriented to study the effects associated with sediment bulk volume concentration on the flow-like regime (i.e., steady state shear condition), whereas the latter was focused on the effects on the yield stress [23] due to the presence of coarse grains.

According to the experiments, the rheological behavior of the mixture are very much influenced by sediment concentration, and in the flow-like regime, it may change from shear thinning to shear thickening, depending on the sediment concentration. It is demonstrated that a generalized Herschel-Bulkley rheological model well represents the flow-like regime of the slurries, being consistent coefficient and power index function of the void ratio of the mixture. Both of them present a limiting value in case of vanishing sediment content and approaching the maximum theoretical sediment concentration, despite the soil characteristics, which affects the fitting parameters. The inclined board experiments put in evidence the role of the sediment concentration and of the coarser grain fraction content on the yield stress.

2. Tested materials and experiments

Materials investigated and experimental methods are widely described in [22]. The investigated materials come from the source area of two real debris flows event occurred in May 1998 (soil B-Montefiorino Irpino) and in March 2005 (soil A-Nocera) in Campania region (southern Italy), which involved the pyroclastic terrains, originated by the volcanic activity of Somma-Vesuvio mount, covering the mountains of that region. Picarelli et al. [24] report an extensive description of their geotechnical characteristics, and several preliminary works have been performed on these materials [25–27].

Figure 1 shows grain size distribution of the collected samples; the soils are sandy silt with a very limited clay fraction. The clay part is slightly plastic though only in the Vesuvian deposits. The gravel part mainly consists of pumices, and secondarily of scoriae and lapilli. The particles are mainly siliceous, and their structure is amorphous and porous (i.e., double porosity system inter- and intra-particle) [24].

The mean physical properties of the sampled soils A and B are specific gravity of soil particles G_S = 2.57, 2.62; dry weight of soil per unit volume γ_d = 7.11, 9.08 kN/m^3; total weight of soil per

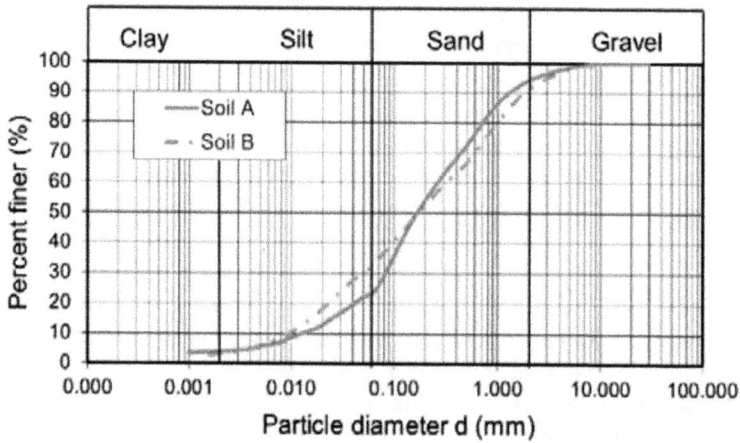

Figure 1. Grain size distribution of the natural soil.

unit volume γ = 12.11, 11.35 kN/m^3; porosity p = 0.71, 0.66; degree of saturation Sr = 0.71, 0.35. Scotto di Santolo et al. [26] reports the extensive description of the geological and geotechnical soils characteristics.

The reconstituted debris flow samples (see **Table 1**) were prepared removing the organic elements and drying out in an oven at 104°C for a day. Then, an appropriate amount of distilled water was used to obtain a soil-water mixture of desired total volumetric concentration Φ_T (ranging from 30 to 42%):

$$\Phi_T = \frac{V_s}{V_s + V_w} \tag{1}$$

where V_S is the volume of solids and V_w the volume of water.

Laboratory activity consists of 22 tests herein reported in **Table 1** as a sack of comprehension. Experiments of Group I refer to fine-grained mixture (having maximum size of grain diameter d = 0.5 mm), whereas Group II and III refer to coarser fraction (i.e., having sediment diameter d > 0.5 mm). The latter are subdivided into four classes: the first two correspond to coarse sand (0.5 mm < d < 1.0 mm) and very coarse sand (1.0 mm < d < 2.0 mm). The latter two (2.0 mm < d < 5.0 mm and 5.0 mm < d < 10.0 mm), are set accounting for the maximum grain size diameter of the collected samples.

The total solid volumetric concentration Φ_T, refers to the bulk volume:

$$\Phi_T = \Phi_f + \Phi_c \tag{2}$$

where Φ_f and Φ_c are the solid volumetric concentration referring to the fine-grained and coarse-grained mixtures, respectively:

$$\Phi_f = \frac{V_{sf}}{V_{sf} + V_{sc} + V_w} \tag{3}$$

$$\Phi_c = \frac{V_{sc}}{V_{sf} + V_{sc} + V_w} \tag{4}$$

In Eqs. (3) and (4), the subscript f, c, w and s refer to fine-grained, coarse-grained materials, water and soil, respectively.

Test	Group	$\Phi_T(\%)$	$\Phi_f{}^\circ(\%)$	$\Phi_c(\%)$				$\Phi_c(\%)$	Soil
				d < 1 mm	d < 2 mm	d < 5 mm	d < 10 mm		
0*	I	32	32					—	A
1*	I	35	35					—	A
2*	I	38	38					—	A
3*	I	40	40					—	A
4*	I	42	42					—	A
5*	I	30	30					—	B
6*	I	32	32					—	B
7*	I	35	35					—	B
8*	I	38	38					—	B
9**	IIa	30	22	8	—	—	—	8	B
10**	IIa	30	17	8	5	—	—	13	B
11**	IIa	30	15	8	5	2	—	15	B
12**	IIa	30	14	8	5	2	1	16	B
13**	IIb	32	24	8	—	—	—	8	B
14**	IIb	32	19	8	5	—	—	13	B
15**	IIb	32	16	8	5	3	—	16	B
16**	IIb	32	15	8	5	3	1	17	B
17**	III	25	25	—	—	—	—	—	B
18**	III	33	25	8	—	—	—	8	B
19**	III	38	25	8	5	—	—	13	B
20**	III	40	25	8	5	2	—	15	B
21**	III	41	25	8	5	2	1	16	B

*Rotational rheometer test.
**Inclined plane test.

Table 1. Experimental program.

2.1. Fine-grained mixture: experimental device and procedure

Fine-grained mixtures tests were performed using a rotational rheometer equipped with vane rotor system. Assuming inertia effects and normal stress differences being negligible, it was possible to derive the shear stress (τ) and the shear rate ($\dot{\gamma}$) from the torque applied to the vane and the angular velocity of the vane rotor, accounting for geometrical device's characteristics [28].

After the complete homogenization of the mixture was ensured, the run-up shear stress ramp started, increasing the applied stress from 0.1 Pa to the maximum stress value (by step of 0.001 Pa). Then, the decreasing shear stress ramp was imposed following the same stress-step, until the initial stress value was applied.

2.2. Coarse-grained mixture: experimental device and procedure

Experiments on coarse-grained were carried out by inclined plane test. It consists of splitting the suspension on the horizontal rough plane in order to obtain a wide layer of material. The tray is progressively inclined until a threshold inclination corresponding to a blatant motion of the mass front, and the experiments were carried on until the full stoppage of the flowing mixture. Accounting for the still, threshold and stoppage condition, and according to the lubrication assumption (i.e., still material thickness much smaller than its longitudinal extent; [29]), it may be assumed a uniform flow condition for the slurry, and momentum balance gives the shear stress distribution within the mixture [22, 29].

2.3. Comparability between sweep test and inclined plane test

The first question is how much the different equipment may give comparable results, and hence if they may be used as alternative method in analyzing rheological behavior. To aim this, inclined plane test on fine-grained mixtures obtained with materials A and B were carried out at the same sediment concentration considered in runs 1–8 (see **Table 1**), and results were compared to those obtained via sweep test. Both in case of rotational rheometer and inclined plane test, the dynamic and static yield stress increase with the solid volumetric concentration and the static yield stress is higher than the dynamic one. Yield stress values obtained from inclined plane are consistent with those resulting from sweep tests, even though the inclined plane test leads to a slight overestimation of the yield stress values according to previous observations [27]. Therefore, inclined plane test may represent a suitable alternative to investigate yielding behavior of dense granular flow mixtures, and it overcomes the shortcoming arising from geometrical limitation of standard rheometer, which confines its operability just to fine-grained slurry (i.e., maximum sediment size smaller than few hundred microns).

3. Flow curves

All the suspensions exhibit a non-Newtonian behavior and the shear stress level increases with the sediment concentration for both soil A and soil B. The experiments show a marked sensitivity of the rheological behavior to granular concentration. **Figures 2** and **3** report stress-strain curve obtained in case of soil A and soil B, whereas **Figures 4** and **5** depict the apparent viscosity ($\eta = \tau/\dot{\gamma}$).

Both yield stress and ultimate apparent viscosity (i.e., viscosity corresponding to the higher stress-strain values) vary over the order of magnitude among the tested solid concentration (ranging from 32 to 42%). The apparent viscosity trend is monotonically decreasing with the shear stress and it tends to a constant value for higher shear stress value. Its values increase, increasing the sediment concentration in both cases of soil A and soil B.

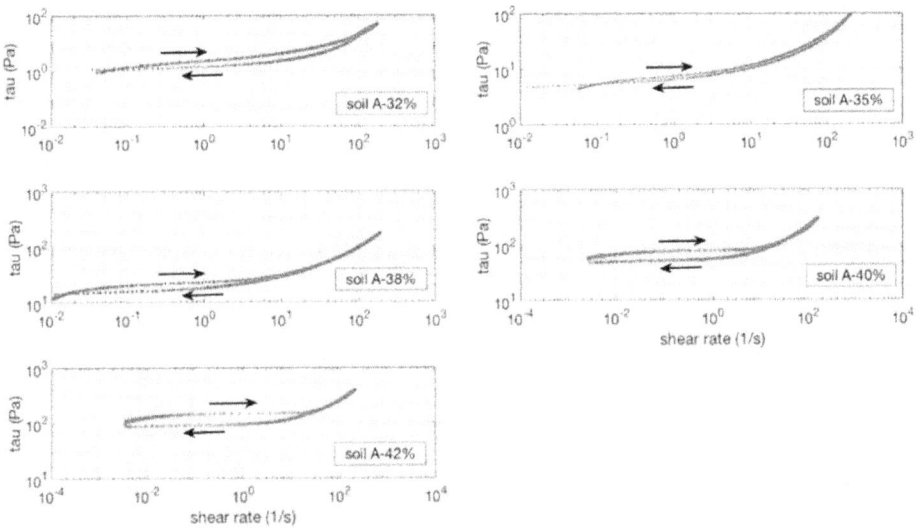

Figure 2. The shear rate $\dot{\gamma}$ versus the shear stress τ for soil A samples. The arrows indicate the increasing-decreasing applied shear stress ramp.

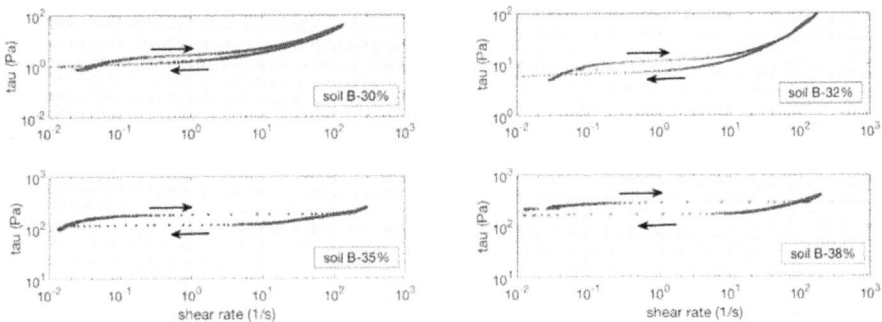

Figure 3. The shear rate $\dot{\gamma}$ versus the shear stress τ for soil B samples. The arrows indicate the increasing-decreasing applied shear stress ramp.

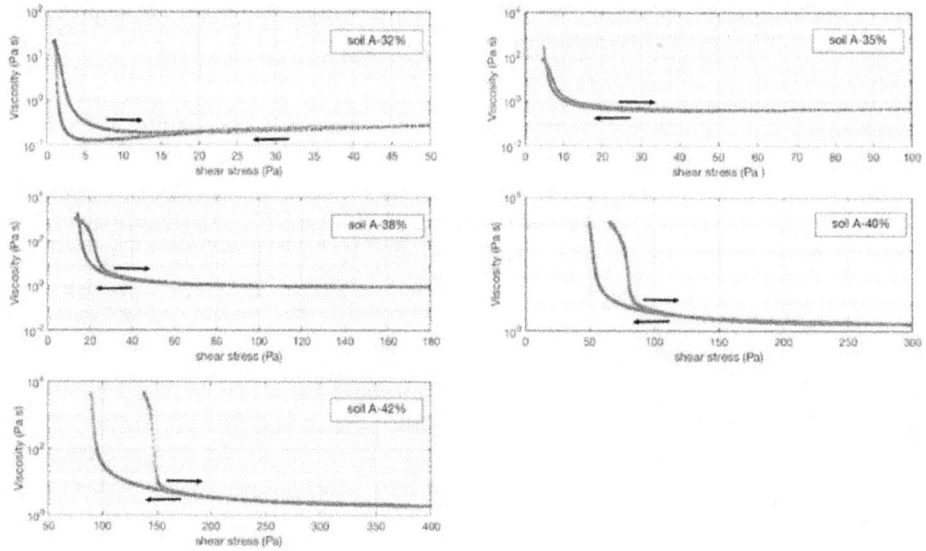

Figure 4. The viscosity η versus τ for soil A samples. The arrows indicate the increasing-decreasing applied shear stress ramp.

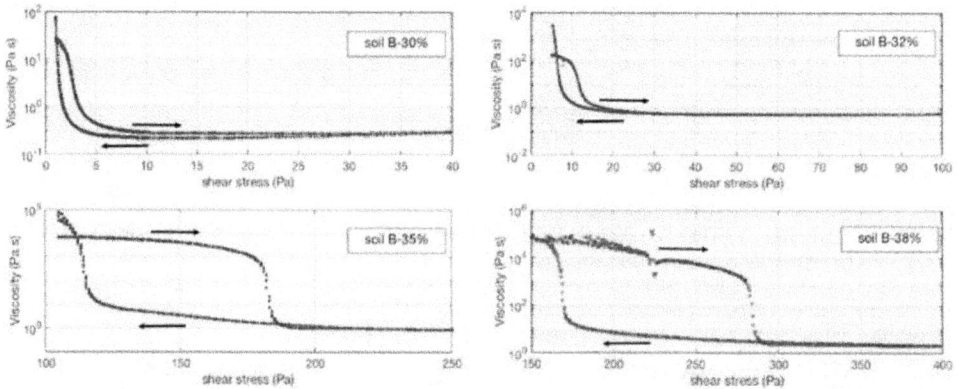

Figure 5. The viscosity η versus τ for soil B samples. The arrows indicate the increasing-decreasing applied shear stress ramp.

It is worth noting thixotropic behavior, as it is evident from the flow curves (**Figures 2, 3**), which exhibit a stress level independent on the shear rate (for shear rate less than 10^0–10^1 s^{-1} depending on sediment concentration) having different values between increasing and decreasing shear stress ramp. The behavior of these granular-fluid mixtures at flow initiation and flow stalling put in evidence that the timescale of microstructure destruction is not the same as that of restructuralization, and it reflects on the yield stress [30]. Notwithstanding the existence of a yield stress, which marks the transition between solid and fluid state, it is still a controversial issue [31], it may be defined a static yield stress τ_{c1}, that is, the critical stress allowing steady state flow (run-up test), and the dynamic yield stress τ_{c2} corresponding to the complete stoppage of the flowing material.

Hysteresis may be better appreciated in **Figure 6** showing a representative sweep test. Increasing the stress level around a critical value (i.e., the static yield stress τ_{c1}), leads to a large increasing of the resulting shear rate, until it reaches the value associated with the end of the stress plateau. It may be considered as a critical value $(\dot{\gamma}_{c1})$, which represents the transition of the material mixture from a yielding to a steady state flow behavior; in fact, no steady flows can be obtained below the critical shear rate [32]. According to the run-down curve, the viscosity remains almost constant over a large range of applied shear stress (**Figures 4** and **5**), since its rate dramatically change in correspondence of the beginning of the stress plateau: it corresponds to the dynamic threshold condition $(\dot{\gamma}_{c2}, \tau_{c2})$.

Referring to dynamic threshold condition, and assuming a representative value for critical shear rate $(\dot{\gamma}_c = 0.1 \text{ s}^{-1})$, the critical shear stress (τ_c) was then estimated from the plateau region in the stress-strain curve by averaging shear stress values, and eventually the dimensionless shear rate and shear stress can be introduced:

$$T = \frac{\tau}{\tau_c} \tag{5}$$

$$G = \frac{\dot{\gamma}}{\dot{\gamma}_c} \tag{6}$$

Dimensionless values are useful to compare different material, despite of total solid concentration. In fact, all the curves collapse to a single one in the range of plateau occurrence

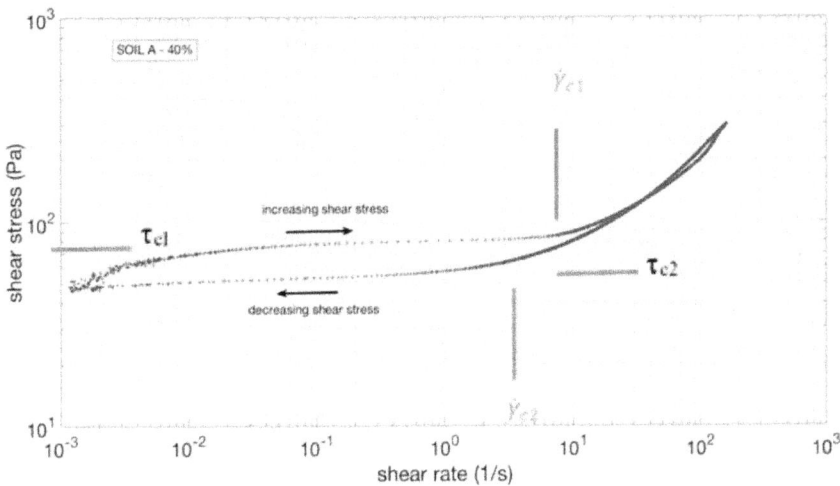

Figure 6. Representative flow curve.

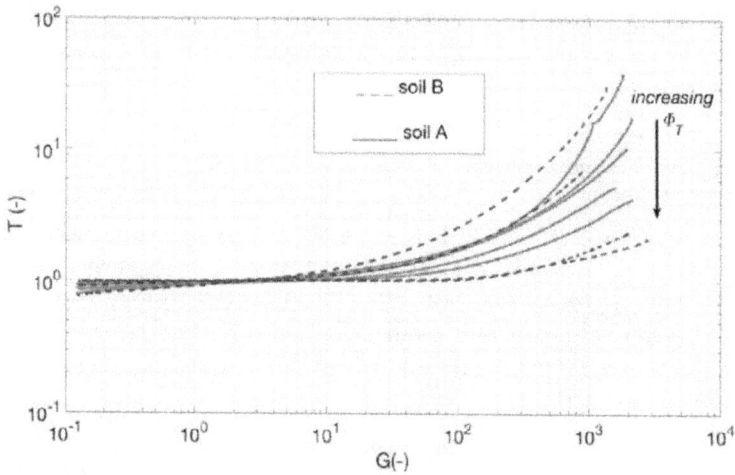

Figure 7. Soil A and soil B. Dimensionless flow curves.

(i.e., around $G = 1$). On the opposite, grain content significantly affects the flow-like regime, and the lower the solid concentration is the higher the stress rate results, independently on the considered soil (**Figure 7**). It is also evident the scatter from the Bingham fluid idealization.

4. Inclined plane results and the effects of grain size distribution on the yield stress

Ancey and Jorrot [20] studied the effect of clay content and concentration of noncolloidal grains on the yield stress, without describing in detail the effects due to granular size distribution, resulting a yield stress model depending on fitting parameters to be extrapolated from experimental results. More recently, Yu et al. [33] performed experimental study on the role of coarse grain in yield stress, and they suggested a yield stress model accounting for an equivalent volumetric solid concentration depending on material characteristics, sediment size, and sediment shape. The model needs some strong approximation, thus it does not present yield stress as a continuous function of sediment concentration. Another aspect related to different grain size refers to particle segregation during flow, which affects the behavior of dense-shared granular flows that are free to dilate [34].

It is generally believed that sediment concentration affects yield stress condition, and shear stress can be expressed as an exponential function of sediment concentration [35–37]. In recent laboratory experiments, Jeong [38] found that little change of silt and sand particles strongly modified the flow behavior, so that increasing sand content, debris flow rheology tends to be more Bingham-like behavior.

It remains still an open question as which are the effects due to grain size distribution on the rheology of debris flows. To this aim, it may be considered the inclined plane test of Group II and III in **Table 1**. Group II tests are mainly oriented to study the effects of increasing the presence of coarse fraction with respect to the total solid volumetric concentration, whereas Group III tests are devoted to stress the effects due to the increasing content of coarse particles, keeping constant the volume of finer grain.

Group II refers to mixtures having constant grain volume concentration (Φ_T = 30%, Group IIa and Φ_T = 32%, Group IIb), varying the relative content of fine and coarse grains. **Figure 8** reports the static and the dynamic yield stress, as a function of total solid volumetric concentration Φ_T and solid volumetric concentration of fine particles Φ_f:

$$\Phi_T/_{\Phi_f} = 1 + \Phi_c/_{\Phi_f} = 1 + V_{sc}/_{V_{sf}} \qquad (7)$$

Run 5 and run 6 related to fine-grained mixtures, are also accounted for, as a reference tests.

Group III consists of mixtures having a constant content of fine particle Φ_f = 25%, and a different concentration of coarse particle Φ_c. **Figure 9** depicts the yield stress value as a function of relative concentration Φ_T/Φ_f.

At constant total solid volumetric concentration Φ_T, the presence of a limited amount of coarse grain leads to a significant reduction on the static and dynamic yield stress, regardless the total solid concentration (see **Figure 8**). Moreover, the less fine grains content is the less yield stress values (both static and dynamic) are, regardless of the coarse particles fraction in the mixtures. The static and the dynamic yield stresses decrease over one order of magnitude if the finer particle content is larger than the coarser grain one. On the opposite, in presence of a comparable content of coarse and fine particle, yield stress slightly varies.

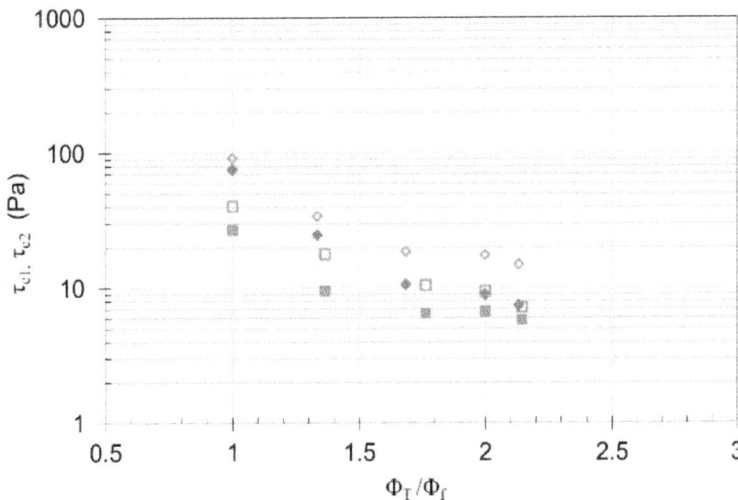

Figure 8. Material B (Φ_T = 30%–red squares, and Φ_T = 32%–blue diamond). Static (empty symbols) and dynamic (filled symbols) yield stress as a function of the ratio between total solid volumetric concentration Φ_T and solid volumetric concentration of fine particles Φ_f (tests #5, #6, and #9–16, see **Table 1**).

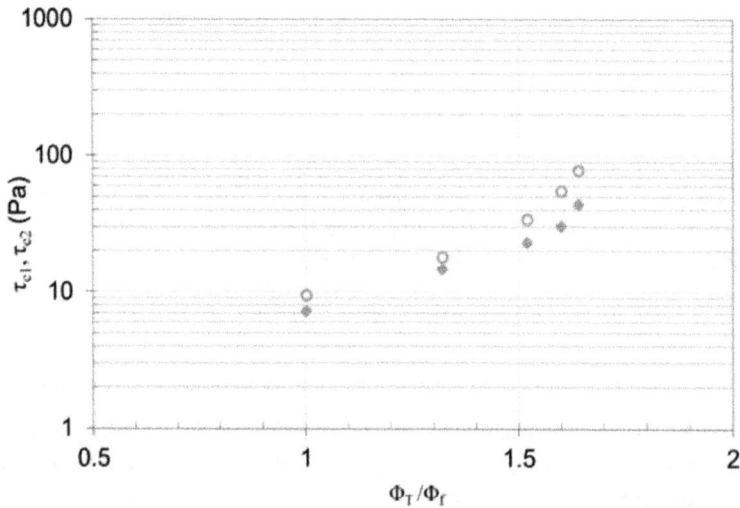

Figure 9. Material B (Φ_f = constant = 25%). Static (empty symbols) and dynamic (filled symbols) yield stress as a function of the ratio between total solid volumetric concentration Φ_T and solid volumetric concentration of fine particles Φ_f (test #5 and tests #17–21, see **Table 1**).

At constant fine particles fraction (**Figure 9**), the increment of coarse grains concentration leads to a significant increase of the yield stress (over one order of magnitude). On the opposite, increasing the volumetric fraction of coarse grains leads to a consistent increasing of the yield stress values.

5. Rheological model

The choice of the most appropriate rheological model is of paramount importance analyzing debris flows, and modeling the runout and deposition fan of slurry flows, which in turn represent the most important aspects in assessing risk associated with geophysical phenomena. A thixotropic flow model may represent both initial structure jamming and aging effects, whereas the non-Newtonian time-independent yield stress model implies the complete reversibility of stress–strain relationship. In many cases, the Herschel-Bulkley model results very similar to the time-dependent thixotropic model [30], and it has widely implemented in viscous-flow simulations [39, 40], even though it still remains challenging the treatment of the non-smoothness constitutive equation [41].

In effect, this chapter focuses on the constitutive equation assuming a simple shearing non-Newtonian flow. Among the varieties of models proposed in literature, the Herschel-Bulkley model seems more appropriate to describe the rheological behavior over the entire range of dynamic condition herein explored, in fact stress–strain rate does not seem linearly proportional in the flow-like regime (see **Figure 7**):

$$\tau = \tau_c + k \cdot \dot{\gamma}^n \tag{8}$$

where k indicates the consistent coefficient, n the power index ($n > 0$ pseudoplastic fluid; $n < 1$; dilatant fluid; $n = 1$ results the Bingham law).

Several other works (e.g., [35, 42]) have already put in evidence that the total solid concentration strongly influences the rheological behavior of granular-fluid mixtures, and it reflects on Herschel-Bulkley generalized model parameters. In the following, the flow-like regime and the yielding condition will be examined separately.

In the range of the stress–strain curve typical of flow-like behavior (i.e., at shear rate greater than 10^0–10^1 s^{-1}), the excess of stress with reference to the yield stress increases with the shear rate, and both parameters k and n result as a function of total sediment concentration (**Figure 10**). Fitting rheological parameters of Herschel-Bulkley model is not trivial. Usually the yield stress τ_c is defined extrapolating the experimental flow curve for vanishing shear rate. In fact, both consistent coefficient k and power index n are very sensitive to the yield stress value [43].

Therefore, it is preferable to split the fitting procedure: first power index n was estimated applying a method proposed by Mullinex [43] no matter of consistent coefficient k or yield stress τ_c. Then, yield stress τ_c was calculated averaging stress value over the plateau region of the flow curve; eventually assuming the already estimated τ_c and n values, parameter k was fitted to the experimental curve (see **Table 2**).

In order to show the influence of grain content on the model's parameter, it is convenient referring to the void ratio (e_0) instead referring to the bulk volume concentration Φ_T:

$$e_0 = \frac{1 - \Phi_T}{\Phi_T} \tag{9}$$

Corresponding to the higher and lower values of void index e_0, the rheological parameters tends to a limiting values (both in case of soil A and B), as it is shown in **Figures 11** and **12**.

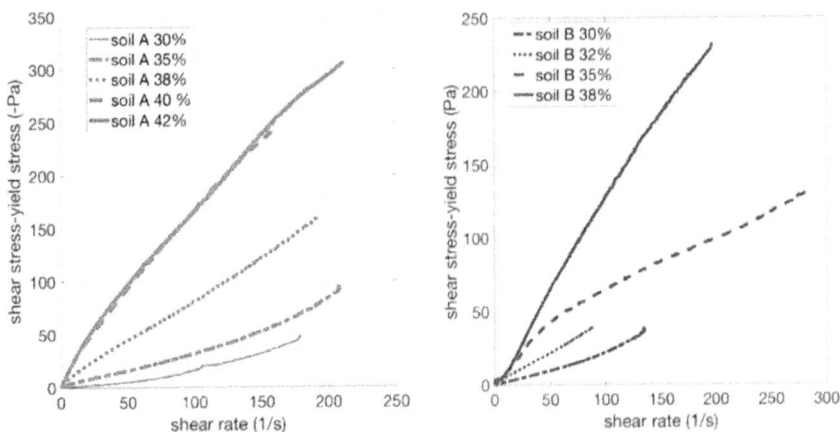

Figure 10. The excess of shear stress with reference to the yield stress versus the shear rate for soil A and soil B tests.

Test	Soil	$\Phi_T(\%)$	$\tau_c(Pa)$	n	$k(Pa\ s^n)$
0	A	32	1.4	1.863	0.003
1	A	35	5.0	1.382	0.055
2	A	38	15.0	0.921	1.236
3	A	40	53.5	0.796	4.236
4	A	42	90.0	0.795	4.526
5	B	30	1.2	1.402	0.036
6	B	32	6.5	1.167	0.212
7	B	35	113.5	0.770	1.700
8	B	38	169.0	0.874	2.294

Herschel-Bulkley rheological parameters.

Table 2. Sweep test on soil A and soil B.

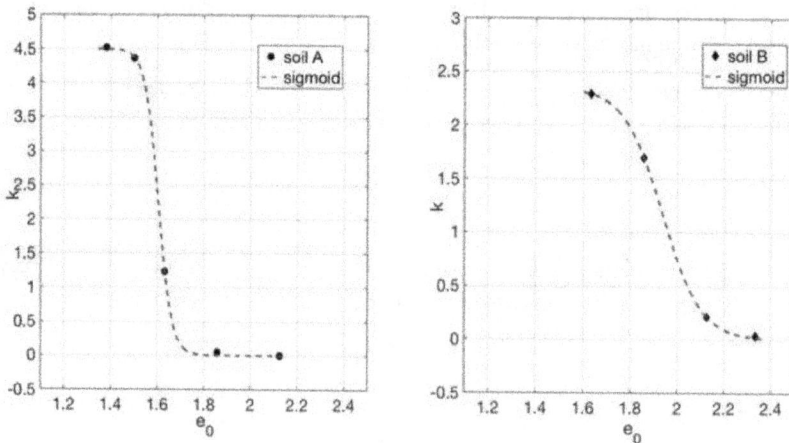

Figure 11. Soil A and B. Consistent coefficient k as a function of void ratio e_0. Sigmoid functions Eq. (10) are plotted according to fitting parameters (see **Table 3**).

These trends suggest considering a sigmoid functions:

$$k(e_0) = \frac{\alpha}{1 + e^{-\lambda(e_0 - \beta)}} + \zeta \tag{10}$$

$$n(e_0) = \frac{a}{1 + e^{-l(e_0 - b)}} + z \tag{11}$$

where $a, b, l, z, \alpha, \beta, \lambda,$ and ζ are fitting parameters depending of mixture characteristics, and their values are shown in **Table 3**.

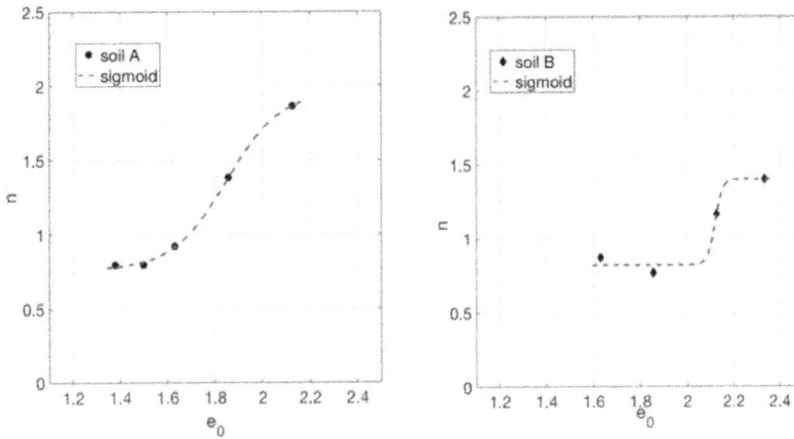

Figure 12. Soil A and B. Power index n as a function of void ratio e_0. Sigmoid functions Eq. (11) are plotted according to fitting parameters (see **Table 3**).

Soil	a	b	l	z	α	β	λ	ζ
A	1.200	1.85	8.808	0.762	4.5	1.603	−34.18	0
B	0.580	2.12	62.79	0.822	2.35	1.936	−12.16	0

Sigmoid functions' parameters Eqs. (10) and (11).

Table 3. Soil A and soil B.

6. Conclusion

Experiments on debris flows reconstituted mixtures analyzed in the present study involves pyroclastic soils presenting a very small clay fraction. They show a behavior of non-Newtonian fluids with yield stress according to several other previous works [15, 22, 44]. Stress–strain curve significantly depart from Bingham idealization, and varying the solid volume concentration, the mixtures show dilatant or pseudoplastic flow behavior, depending on the granular concentration, no matter of the considered soil characteristics. According to other studies [35], the solid content greatly affects the behavior of these mixtures during the flow, as it is evident studying the influence of the solid volumetric concentration on the rheological parameters of the mixtures.

Accounting for the run out and stoppage phase of debris flow, we may refer to the dynamic yield stress and to a simple shearing regime (i.e., simple yield stress fluid) [45]. Under this assumption, Herchel-Bulkley model reasonably applies to the experimental flow curve. The power index varies in the range $n = [0.87{-}1.86]$ over the whole set of experiments; it shows a dilatant fluid behavior for the lower grain volume concentration, and progressively tends to a shear thinning fluid increasing the sediment content. Analogously consistent coefficient k [0.003–4.526] varies with sediment concentration or sediment void ratio, which seems more appropriate in order to define functional relationships.

Both parameters show a limiting value corresponding to the higher and the lower grain content. Therefore, the proposed rheological model applies a sigmoid function for both consistent coefficient and power index, whose fitting coefficients depend on mixture characteristic.

The inclined plane experiments at constant sediment bulk volume, show that increasing the relative content of coarser fraction leads to a diminishing of yield stress. On the other hand, keeping constant the finer sediment content, the more relevant coarse fraction is the higher yield stress results.

The relative concentration of coarse and fine particle seems to discriminant the rheological behavior. In the presence of dominant fine grain fraction, slight increase of the coarse grain fraction leads to a dramatic decrease of both static and dynamic yield stress values. When the concentration of coarse particles in the mixture increases and become similar to that of fine particles, the values of the rheological parameters more slightly decrease.

Acknowledgements

I gratefully thank to Dr. Anna Maria Pellegrino for the experimental data.

Conflict of interest

The authors declare no conflict of interest.

Author details

Leonardo Schippa

Address all correspondence to: leonardo.schippa@unife.it

Department of Engineering, University of Ferrara, Ferrara, Italy

References

[1] Forterre Y, Poliquen O. Granular flows. In: Proceedings of Séminaire Poincarré XIII. Paris, France: Institut Henry Poincarre; 2009. pp. 69-100. Available online: http://www.bourbaphy.fr/pouliquen.pdf [Accessed: 1 November 2017]

[2] Takahashi T. Debris Flows. IAHR Monograph Series. Rotterdam: Balkema; 1991

[3] Coussot P, Laigle D, Arratano M, Deganutti A, Marchi L. Direct determination of rheological characteristics of debris flow. Journal of Hydraulic Engineering. 1998;**124**:865-868

[4] Ancey C. Role of particle network in concentrated mud suspensions. In: Rickenmann D, Chen C, editors. Debris-Flow Hazards Mitigation: Mechanics, Prediction, and Assessment. Millpress Science Publishers; 2003. pp. 257-268

[5] Schatzmann M, Bezzola GR, Minor HE, Windhab EJ, Fischer P. Rheometry for large particulated fluids: Analysis of the ball measuring system and comparison to debris flow rheometry. Rheologica Acta. 2009;**48**:715-733

[6] Scotto di Santolo A, Pellegrino AM, Evangelista A, Coussot P. Rheological behaviour of reconstituted pyroclastic debris flow. Geotechnique. 2012;**62**:19-27

[7] Pellegrino AM, Schippa L. Macro viscous regime of natural dense granular mixtures. International Journal of Geomate. 2013;**4**:482-489

[8] Pellegrino AM, Schippa L. Experimental analysis for fine-grained and large-grained soils involved in debris flow at the solid-fluid transition. In: Fukuoka S, Nakagawa H, Sumi T, Zang H, editors. Advances in River Sediment Research. London: CRC Press Taylor & Francis Group; 2013. p. 39

[9] Iverson RM. The debris-flow rheology myth. In: Rieckenmann D, Lung Chen C, editors. Proceedings of the International Conference on Debris-Flow Hazards Mitigation: Mechanics, Prediction, and Assessment; 10–12 September 2003; Davos, Switzerland. Davos, Switzerland: Mills Press; 2003. pp. 303-314. Available Online: http://www.scopus.com/inward/record.url?eid=2-s2.0-10644228696&partnerID=tZOtx3y1 [Accessed: 1 September 2017]

[10] Ancey C. Plasticity and geophysical flows: A review. Journal of Non-Newtonian Fluid Mechanics. 2007;**142**:4-35

[11] Schippa L. Two-phase model for plane shear of dense granular flow. Journal of Hydrology and Hydromechanics. 2000;**48**:334-355

[12] MiDi G. On dense granular flows. European Physical Journal E: Soft Matter and Biological Physics. 2004;**14**:341-365

[13] Stickel JJ, Powell RL. Fluid mechanics and rheology of dense suspensions. Annual Review of Fluid Mechanics. 2005;**37**:129-149

[14] Ancey C, Coussot P, Evesque P. A theoretical framework for granular suspensions in a steady simple shear flow. Journal of Rheology. 1999;**43**:1673-1699

[15] Coussot P. Mudflow Rheology and Dynamics. Rotterdam The Nederlands: Balkema; 1997

[16] Chen H, Lee. Runout analysis of slurry flows with Bingham model. Journal of Geotechnical and Geoenvironmental Engineering. 2002;**128**(12):1032-1042

[17] Iverson RM. The physics of debris-flow. Reviews of Geophysics. 1997;**35**(3):245-296. DOI: 10.1029/97RG00426

[18] Sengun MZ, Probstein RF. Bimodal model of slurry viscosity with applications to coal slurries. Part 1. Theory and experiment. Rheologica Acta. 1983;**28**:382-393

[19] Coussot P, Piau JM. A large-scaled field concentric cylinder rheometer for the study of the rheology of natural coarse suspensions. Journal of Rheology. 1995;**39**:105-124

[20] Ancey C, Jorrot H. Yield stress for particle suspensions within a clay dispersion. Journal of Rheology. 2001;**45**:297-319

[21] Coussot P. Yield stress fluid flows: A review of experimental data. Journal of Non-Newtonian Fluid Mechanics. 2014;**211**:31-49

[22] Pellegrino AM, Schippa L. A laboratory experience on the effect of grains concentration and coarse sediment on the rheology of natural debris-flows. Environment and Earth Science. (Submitted)

[23] Malet JP, Remaître A, Maquaire O, Ancey C, Locat J. Flow susceptibility of heterogeneous marly formations: Implications for torrent hazard control in the Barcelonnette Basin (Alpes-de-haute-Provence, France). In: Rickenman D, Chen C, editors. Debris-Flow Hazards Mitigation: Mechanics, Prediction, and Assessment. Rotterdam: Millpress; 2003. pp. 351-362

[24] Picarelli L, Evangelista A, Rolandi G, Paone A, Nicotera MV, Olivares L, Scotto di Santolo A, Lampitiello S, Rolandi M. Mechanical properties of pyroclastic soils in Campania region. In: Tan TS, Phoon KK, Hight DW, Leroueil S, editors. Characterization and Engineering Properties of Natural Soils. London: CRC Press Taylor & Francis Group; 2007. pp. 2331-2338

[25] Scotto di Santolo A, Pellegrino AM, Evangelista A. Experimental study on the rheological behaviour of debris flow. Natural Hazards and Earth System Sciences. 2010;**10**:2507-2514

[26] Scotto di Santolo A, Pellegrino AM, Evangelista A, Coussot P. Rheological behaviour of reconstituted pyroclastic debris flow. Geotechnique. 2012;**62**:19-27

[27] Pellegrino AM, Schippa L. Macro viscous regime of natural dense granular mixtures. International Journal of Geomate. 2013;**4**:482-489

[28] Nguyen QD, Boger DV. Direct yield stress measurement with the vane method. Journal of Rheology. 1985;**29**:335-347

[29] Coussot P. Rheometry of Pastes, Suspensions and Granular Materials: Application in Industry and Environment. New York: John Wiley and Sons Inc. Publications; 2005

[30] Jeon CH, Hodges BR. Comparing thixotropic and Herschel–Bulkley parameterizations for continuum models of avalanches and subaqueous debris flows. Natural Hazards and Earth System Sciences. 2018;**18**:303-319

[31] Bonn D, Denn MM, Berthier L, Divoux T, Manneville S. Yield stress materials in soft condensed matter. Reviews of Modern Physics. 2017;**89**(3):035005(40)

[32] Ovarlez G, Rodts S, Chateau X, Coussot P. Phenomenology and physical origin of shear localization of the shear banding in complex fluids. Rheologica Acta. 2009;**4**:831-844

[33] Yu B, Chen Y, Liu Q. Experimental study on the influences of coarse particle on the yield stress of debris flow. Applied Rheology. 2016;**26**(42997):1-13

[34] Gray JMNT. Particle segregation in dense granular flows. Annual Review of Fluid Mechanics. 2018;**50**:407-433

[35] Major JJ, Pierson TC. Debris flow rheology: Experimental analysis of fine-grained slurries. Water Resources Research. 1992;**28**:841-857

[36] Philips CJ, Davies TRH. Determining rheological parameters of debris flow materials. Geomophology. 1991;**14**:573-587

[37] Kaitna R, Rickenmann D, Schatzmann M. Experimental study on the rheological behaviour of debris-flow material. Acta Geotechnica. 2007;**2**:71-85

[38] Jeong WJ. Grain size dependent rheology on the mobility of debris flows. Geosciences Journal. 2010;(4):359-369

[39] Fornes P, Bihs H, Nordal S. Implementation of non-newtonian rheology for granular flow simulation. In: Proceedings of 9th National Conference on Computational Mechanics MekIT'17; 11–12 May 2017; Trondheim, Norwey

[40] Remaître A, Malet JP, Maquaire O, Ancey C, Locat J. Flow behaviour and runout modelling of a complex debris flow in a clay-shale basin. Earth Surface Processes and Landforms. 2005;**30**:479-488

[41] Saramito P, Wachs A. Progress in numerical simulation of yield stress fluid flows. Rheologica Acta. 2017;**56**:211-230

[42] O'Brien JS, Julien PY. Laboratory analysis of mud flow properties. Journal of Hydraulic Engineering. 1988;**114**(8):877-887

[43] Mullineux G. Non linear least squares fitting of coefficients in the Herschel Bulkley model. Applied Mathematical Modelling. 2008;**32**:2538-2551

[44] Schatzmann M, Bezzola GR, Minor H, Windhab EJ, Fischer P. Rheometry for large particulated fluids: Analysis of the ball measuring system and comparison to debris flow rheometry. Rheologica Acta. 2009;**48**:715-733

[45] Ovarlez G, Cohen-Addad S, Krishan K, Goyon J, Coussot P. On the existence of a simple yield stress fluid behavior. Journal of Non-Newtonian Fluid Mechanics. 2013;**193**:68-79

Inside the Phenomenological Aspects of Wet Granulation: Role of Process Parameters

Veronica De Simone, Diego Caccavo,
Annalisa Dalmoro, Gaetano Lamberti,
Matteo d'Amore and Anna Angela Barba

Additional information is available at the end of the chapter

http://dx.doi.org/10.5772/intechopen.79840

Abstract

Granulation is a size-enlargement process by which small particles are bonded, by means of various techniques, in coherent and stable masses (granules), in which the original particles are still identifiable. In wet granulation processes, the powder particles are aggregated through the use of a liquid phase called binder. The main purposes of size-enlargement process of a powder or mixture of powders are to improve technological properties and/or to realize suitable forms of commercial products. A modern and rational approach in the production of granular structures with tailored features (in terms of size and size distribution, flowability, mechanical and release properties, etc.) requires a deep understanding of phenomena involved during granules formation. By this knowledge, suitable predictive tools can be developed with the aim to choose right process conditions to be used in developing new formulations by avoiding or reducing costs for new tests. In this chapter, after introductive notes on granulation process, the phenomenological aspects involved in the formation of the granules with respect to the main process parameters are presented by experimental demonstration. Possible mathematical approaches in the granulation process description are also presented and the one involving the population mass balances equations is detailed.

Keywords: granular materials, wet granulation process parameters, granule growing, granule breakage, granulation mathematical description

1. Introduction

Granulation, also known as agglomeration, pelletization, or balling, is a "size-enlargement process" of small particles into larger coherent and stable masses (granules), in which the original particles are still identifiable [1]. The aim of the granulation process is to improve the properties of the final product compared to the powder form, such as giving better flow properties for safer and cheaper transport and storage, lowering of caking and lump formation (especially for hygroscopic materials), improving heat transfer features, obtaining a more uniform distribution of the active molecule, lowering powder dispersion in the environment, linked to a reduced inhalation, handling and explosion risks hazard [1, 2]. Granules have, therefore, received a great interest in many industrial fields, from mineral processing to agricultural products, detergents, pharmaceuticals, foodstuffs, nutraceuticals, cosmetics and zootechnical products [3].

In pharmaceutical field, solid dosage forms remain an important part of the overall drug market, despite the success and the development of new pharmaceutical forms. The oral solid dosage forms market was of $571 million in 2011 and projected to reach $870 million at the end of 2018 [4]. In particular, among novel drugs approved by FDA, 46% in 2014 and 32% in 2016 were solid dosage products [5], most of them made of granules. The most important pharmaceutical industries, such as Patheon, Aesica, Rottendorf Pharma GmbH, Catalent Pharma, continually do investments in oral solid manufacturing solutions, including the development of granulation processing methods [6]. Granulated products are also highly used in the fertilizers field: about 90% of fertilizers are applied as solids, less as powder, more in granular form. The global demand for fertilizer nutrients was estimated to be 184.02 million tons in 2015, and it is forecast to reach 201.66 million tons by the end of 2020 [7]. The animal feed additives global market was estimated at 256.8 kilo tons in 2015, and in particular, industries aim to develop new technologies [8], very often based on granulation principles, to provide stabilization and effective protection of the active components in the finished products [9].

In spite of its great importance and over 40 years of research, granulation process is still based on practical experience, i.e., there is a qualitative understanding of both the granule growth mechanisms and the effects of different variables on agglomeration phenomena. This constitutes a great problem for industries exploiting many and frequently changing formulations with widely varying properties (e.g., food, pharmaceuticals and agricultural chemicals). Thus, new formulations always need expensive and lengthy laboratory and pilot-scale testing. Moreover, even when pilot-scale testing is ok, there is still a significant failure rate during scale up to the industrial production [10]. Over the past decade, however, design, scale-up, and operation of granulation processes have been considered as quantitative engineering and significant advances have been made to quantify the granulation processes [3]. Firstly, granulation must be recognized as an example of "powder particle design": granule final features are controlled by a perfect combination of formulation design, i.e., choice of the feed-material properties, and process design, i.e., choice of equipment type and operating conditions [11, 12]. The granule final properties are also determined by the interaction of phenomena coexisting in the granulation process simultaneously [1]: physical transformations of the powder particles with relevant kinetic mechanisms (wetting and nucleation, consolidation and growth, and breakage and attrition) and aggregation rate are controlled by operating conditions and feeding material properties [3]. Therefore, for the quantitative analysis of the granulation process, both a careful characterization of the feed-material properties and knowledge of operating parameters and phenomenological aspects are needed [3, 13]. Basically, the granulation methods are divided into dry and wet ones. Dry granulation is based on the use of mechanical compression (slugs) or compaction (roller compaction), while the wet granulation exploits a granulation liquid phase called "binder" to agglomerate powder particles by formation of a wet mass by adhesion. This second process involves a final step of granules stabilization by removal of the wetting phase to make the relevant bonds permanent. Among these two techniques, wet granulation is the most widely used [2, 10, 14]. To produce granular structures with tailored features (in terms of size and size distribution, flowability, mechanical and release properties, etc…), a deep understanding of phenomena involved during granule formation is required. This can also lead to optimized processes in terms of production costs and other types of resources involved.

The purpose of this chapter is to emphasize the phenomenological aspects involved in the formation of the granules with respect to the main parameters of the wet granulation process. To this aim, experimental demonstrative campaigns of wet granulation runs were performed by changing several formulations and process parameters to detail the effect of each single parameter both on the granulation stages and on the obtained granular materials features. Mathematical modeling approaches' description of granulation process was then presented. Usually, in literature, the approach to the granulation process is based on either experimental tests (to study granules properties) or on modeling studies. The novelty of this work is the integrated approach exploiting both experimental studies to understand the involved basic phenomena in powders aggregation, with the aim to optimize the process, and modeling aspects to verify and predict the experimental results.

2. Wet granulation process: apparatus and process parameters

2.1. Apparatus features

As briefly introduced, the wet granulation technique allows the formation of granules through the addition of a binding phase to a powder bed. It is usually performed in four steps: (1) homogenization of dry powders, (2) wetting by binder addition, (3) wet massing when liquid feeding system switches off, and (4) drying of the finished product [15]. The most common apparatuses used for granules production are: tumbling granulators, both batch and continuous low and high shear mixers, fluid bed granulators [16]. In tumbling granulators (including discs, drums, pans, and similar equipment), the particles motion is assured by the tumbling action caused by the balance between gravity and centrifugal forces. In particular, the powder feed is fed to the disc, typically at the edge of the rotating granular bed, and the binder is added through a series of nozzles distributed across the face of the bed. Discs and drums generally operate continuously and have large throughputs; thus, they are extensively used in mineral processing and fertilizer granulation [17]. Low and high shear mixers are mechanically agitated containers that promote an efficient mixing, especially of cohesive materials. Such mixers exert intense local shear force actions on the powder, which break the small cohesive aggregates, promoting good dispersion of the liquid and effective consolidation of the product [15]. In fluid bed granulators, the powder bed is first fluidized by a flow of air injected upward through a distributor plate at the base of the granulator, and then the liquid binder is sprayed through a nozzle onto the fluidized bed to agglomerate powder in granules. When binder spraying is stopped, the granules continue to dry in the fluidizing airstream, avoiding the use of a following drying step [18]. This type of granulator is flexible, relatively easy to scale, difficult for cohesive powders, and good for coating applications [16]. Wet granulation has also witnessed various technical and technological innovations such as steam granulation, moist granulation, thermal adhesion granulation, melt granulation, freeze granulation, foamed binder granulation, and reverse wet granulation. For example, steam granulation exploits water steam as binder, providing a more rapid diffusion into the powder and a more favorable thermal balance during the drying step [14].

In general, three fundamental sets of rate processes determine wet granulation behavior: (1) wetting and nucleation, (2) consolidation and growth, (3) breakage and attrition (see schematization in **Figure 1**). Wetting/nucleation is the initial step where the liquid binder comes in contact with the dry powder bed causing the adhesion among particles to obtain a distribution of small aggregates (nuclei). During the consolidation and growth phase, the particles collide in the granulator and the nuclei begin to grow (particles increase in size and volume) for the deposition of additional material on the nuclei surface. Finally, attrition and breakage phase is characterized by the rupture of granules with relevant formation of small particles, due to both the impacts in the granulator and product handling. These mechanisms coexist in all the wet granulation processes, even if their importance is related to the process type. For example, in the fluid-bed granulation, the wetting phase prevails while in the high-shear

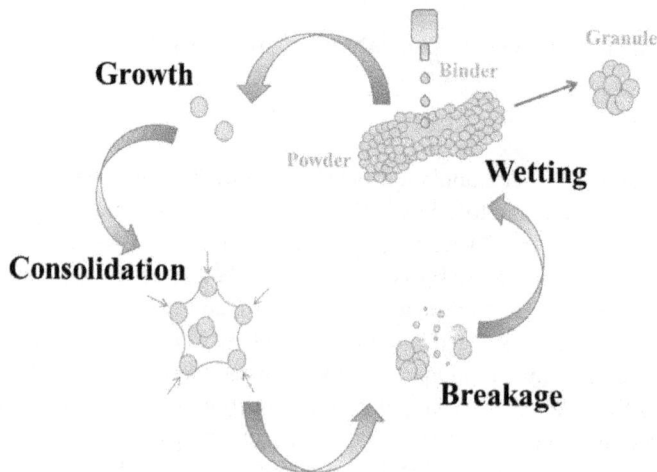

Figure 1. Mechanisms involved in wet granulation.

granulation, the consolidation step is predominant. To estimate what will be the characteristics of the granules, it is necessary to know that each of the phases presented has a fundamental role that must be predictable. In fact, once these processes have been analyzed, it is possible to predict, at least theoretically, the type of equipment and operating conditions to be used to obtain a good granulation [2].

2.2. Effect of process parameters

Several parameters can play a fundamental role on the basic mechanisms of wet granulation and therefore on the product final properties [19]. In particular, granule features depend on ingredients formulation (binder and powder properties and their interaction and proportion), process, and equipment parameters, these last two depending on the kind of the used apparatus. Thus, both material and operating variables together define the kinetic mechanisms and rate constants of wetting, growth, consolidation, and attrition [16].

2.2.1. Formulation parameters

2.2.1.1. Effect of binder addition rate

Due to the use of a liquid phase as binder, the ratio between the liquid and solid phases would affect the granule final properties [20]. If the liquid/solid ratio increases (due to a high amount of used liquid), nucleation is favored, but at the same time, it is possible that overwetting phenomena may occur, with consequent formation of a mixture and not granules. Moreover, as the quantity of the added liquid increases, the granule saturation, i.e., the ratio between the liquid volume and the interstitial granule volume, changes. However, the addition of too much liquid implies a larger granule size because of a high saturation; in the same way, a low saturation does not allow the granules to growth. The needed amount of liquid must be increased as the size of the powder particles decreases [19, 21–23].

2.2.1.2. Effect of binder delivery method

The addition of the wetting phase can take place in three different ways, i.e., by pouring, by spraying, or by making it melt, and it is closely linked to the nucleation regime, which has in turn a substantial effect on the product final features [24, 25]. A uniform liquid spray with small droplets size will have the greatest coverage throughout the powder bed and will prevent localized overwetting of the granules, which can result in oversized particles [16]. Moreover, both when pouring and when spraying, the particles size distribution (PSD) is initially bimodal and it tends to be unimodal for high granulation times [26]. In melting technique, however, the obtained granules will be less coarse and only for high granulation times a bimodal distribution will develop [19, 27].

2.2.1.3. Effect of binder properties

Binder viscosity and surface tension are the properties, which more influence the granulation process because the collision energy necessary to agglomerate particles depends on them [28–31]. In particular, a higher binder solution viscosity could lead to larger granule size and less needed binder amount to start the granule growth in both high-shear [32, 33] and fluid-bed [34] granulation processes. This is due to the fact that a high liquid viscosity requires more energy to break up the liquid droplets (less binder spreadability); hence, larger droplets are formed, which consequently give larger granules. However, at too high viscosity, droplets will be unable to spread throughout the bed causing the reduction in collisions and relevant growth [35]. Surface tension and capillary forces always act to pull particles together, and their magnitudes depend on the liquid bridge formed between the particles [36]. Reducing binder surface tension causes the decrease in the capillary suction pressure and friction resistance, leading therefore to an improved wettability and spreading efficiency [10]. Moreover, also the solvent used for the formulation of the binder solution (only water, alcoholic, or hydroalcoholic solutions are usually used) could significantly change granule properties for its impact on binder wettability and spreadability [28].

2.2.1.4. Effect of powder particles' size and solubility

Powder particles' size influences the amount of binder to be used: a larger liquid amount is required to establish liquid bridges between the powders of lower size, thus with a high surface area [37, 38]. Moreover, the high surface area allows the availability of more contact points between colliding particles bringing as final result to stronger granules, which, however, have a more porous structure [10, 29, 30]. Perhaps the larger surface area allows also a higher growth tendency of the smaller particle fraction probably due to a more efficient nucleation and coalescence [39]. Moreover, a larger solubility of the solid excipient in the granulating solvent is able to decrease the solvent amount needed for granule formation, and granules with uniform particle size distribution and a reduced friability will be formed [28].

2.2.2. Equipment and process parameters

In general, equipment and process variables impact on mixing, agglomerating, and drying operations. For example, equipment variables in fluid bed granulators are related to the apparatus design with the aim to fluidize, thus granulate and dry the product. Therefore, air distribution plate must be appropriately designed depending on the powder properties. For example, a product with low bulk density will require a low fluidizing velocity, thus a distributor plate having a small open area. Moreover, a blower with appropriate pressure drop will fluidize the process material adequately. A proper use and cleaning of filters to retain entrained particles also must be carried out in fluidized bed [16]. However, for both mixers and fluid beds, the bowl geometry is also considered a factor with large impact on the agglomeration process. Moreover, for low and high shear mixers, the impeller and chopper design affects the flow patterns and powder flow dynamics in the bowl, by varying the volume of powder mixture swept out by the impeller itself: a high-swept volume provokes high densification of the agglomerates and narrow granule size distributions [40].

Process variables in fluid bed granulators agglomeration is highly dependent on: process inlet air temperature, atomization air pressure, fluidization air velocity and volume, liquid spray rate, nozzle position and number of spray heads, and product and exhaust air temperature [41]. Inlet-process air temperature depends on both the binder type and the heat sensitivity of powder bed. For example, higher temperatures will cause binder faster evaporation with the relevant production of smaller and friable granules [42]. Process variables in high/low-shear mixers are essentially related to the impeller and chopper relative speed, granulating solution addition rate, both global granulation and wet-massing time [43]. In general, we can conclude that for all types of apparatuses, both equipment and process parameters define the two most important operating conditions in wet granulation process, i.e., mixing performance and residence time.

3. Inside wet granulation runs: experimental evidences

The impact that the aforementioned process parameters have on final granule properties is continuously studied to predict the final product quality. A possible approach to study the wet granulation process was that of a recent study conducted by [44], devoted to first planning experiments by the design of experiments (DoEs) and then to use the results to give correlations between product properties and process parameters. In particular, the central composite design (CCD) statistical protocol was applied for planning the experimental campaign about the production of hydroxypropyl methylcellulose (HPMC 20, Pentachem Srl, San Clemente RN-Italy) granules with distilled water as the binder phase and using a bench scale low-shear granulator apparatus. In brief, a given amount of HPMC 20 powder was placed in the low-shear granulator, and then the addition of the binder phase was carried out by spraying it by an ultrasonic device. The produced granulates were stabilized by dynamic drying, using hot air (65°C) for 1 h, collected, and then separated by a manual sieving with cut-off sizes as follows: 2 mm, 0.45 mm, and a metal collection pan. Three particles fractions were obtained: a fraction of "big scrap," i.e., particles with size larger than 2 mm, a fraction of "small scrap," i.e., particles with size smaller than 0.45 mm, and a fraction of "useful," i.e., particles with size between 0.45–2 mm. The range size 0.45–2 mm was considered as the fraction of interest being a size typical range of commercial granulated food, pharmaceutical, and zootechnical products. Finally, only the fraction of useful was subjected to characterization protocols carried out by adopting the American Society for Testing and Materials (ASTM) standards.

Firstly, screening experiments were performed in order to determine the independent variables and their interactions playing a significant role on the dependent variables, i.e., on the final features of granulated product, such as granulation yield (defined as mass percentage of dry granules with size between 0.45 and 2 mm) and flowability properties (Carr Index, Hausner Ratio and Angle of Repose). The screening work showed that some parameters can be fixed, such as the powder mass (50 g) and the process time of 20 min (by observing that longer times caused granule-breaking phenomena). Instead, the impeller rotation speed, the binder volume at constant mass, and the binder flow rate were the parameters (factors) with larger influence on the final granule properties. The factors have a values limited range: high binder volume (greater than 100 ml) or high binder flow rate (greater than 58 ml/min) involved overwetting phenomena, low amounts of liquid (lesser than 50 ml) did not form granules, high impeller rotation speed (greater than 112 rpm) generated solid particulate breaking. At this point, for each factor, three intensities (levels) were used, i.e., the minimum, medium, and maximum values, and combined by the CCD method. The performed runs have underlined that there are process operating conditions, which combined together can produce granules with size smaller than the requested one, i.e., failure of the aggregation phenomena (**Figure 2A**).

Others, instead, can achieve clusters of powder and binder, i.e., overwetting phenomena, that is a condition to avoid (**Figure 2C**). The best conditions of granulation, able to produce granules with a defined size (0.45–2 mm) and good flowability together with a high granulation process yield, were obtained by working with a high impeller rotation speed, i.e., 112 rpm, a high binder volume, i.e., 100 ml, and a low binder flow rate, i.e., 17 ml/min (**Figure 2B**). Then, semiempirical correlations between granule properties and process parameters were developed by describing the experimental data with several model equations. Akaike information criterion and R-square calculations showed that the best comparison between experimental data and model predicted values was attained by using the second-order polynomial equation. The proposed correlations were then validated by new granulation tests, not included in the work plan, underlining their ability to predict the granule final properties in terms of flowability and granulation yield. It is important to note that several studies in the literature describe the correlations between process parameters and granule properties; however, they use different apparatuses and final products (tablets). It was the first time that, for such similar granulation systems, semiempirical correlations were able to give the combined effect of impeller rotation speed, binder volume, and binder flow rate on granulation yield and flowability [44].

The found combination of process parameters (for granules better final properties) is thus used in the production of loaded granules, with the aim to evaluate the effect of two formulation variables, molecule solubility and binder type, on their physical, mechanical, and release properties. First of all, the best loading method for a hydrophilic molecule, vitamin B12, in HPMC granules was investigated. Vitamin B12 was incorporated in the HPMC granules by two different loading methods: according to the method 1, it was dissolved in the liquid binder phase (here, the binder was a solution of water and vitamin B12); according to the method 2, the vitamin B12 was premixed with HPMC powders at an impeller rotation rate of 78 rpm for 10 min. It was observed that the loading method did not have significant effects on either the

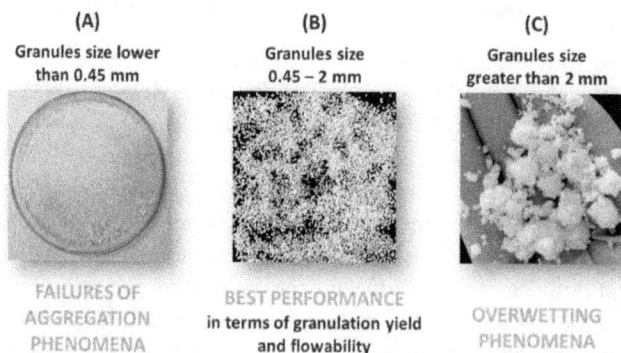

(A)	(B)	(C)
Granules size lower than 0.45 mm	Granules size 0.45 – 2 mm	Granules size greater than 2 mm
FAILURES OF AGGREGATION PHENOMENA	BEST PERFORMANCE in terms of granulation yield and flowability	OVERWETTING PHENOMENA

Figure 2. Granules obtained with different combination of factors and levels: (A) particles with size smaller than 0.45 mm, (B) particles with size between 0.45 and 2 mm, that is the size range required, and (C) particles with size larger than 2 mm.

granules flowability or the yield, but a better dispersion of vitamin B12 inside the HPMC polymer matrix was achieved by the method 1, perhaps thanks to the high solubility of this vitamin in the binder and the relevant uniform spray by the ultrasonic atomization device [45]. Thus, by exploiting the most successful method 1, three different payloads of B12 (1, 2.3, and 5% w/w) were tested. It was observed that a high vitamin load (5%) reduced the granulation yield and brought to the formation of more rigid granules if compared to unloaded ones and those loaded at 1 and 2.3% of B12. Vitamin release kinetics was slower when it was added with a 1% load, thus suggesting a better incorporation. Moreover, by comparing kinetics of fresh and 1 month-aged granules, they showed similar trends, thus no effect of the storage on release properties of loaded granules was observed [45].

Moreover, the effect of the incorporation of a lipophilic vitamin was also tested by using the loading method 1. Due to its lipophilic properties, a binder composed of a solution made by distilled water and ethanol was used. Results showed that the use of ethanol gave a reduced granulation yield and granules with less defined shape, smaller dimensions, more friable structure, worse flowability, and slightly faster polymer erosion. It was demonstrated that the molecule solubility did not affect either granules' physical or mechanical properties, but it had effect on the molecule release mechanism (diffusion for the hydrophilic molecule and erosion for the lipophilic one) [46].

A deep understanding of the phenomena involved during the wet granulation process can lead to optimized processes that obtain the maximum benefits without increasing the production costs. In light of that, studies on the evolution of particle size distribution (PSD) during the granulation process were performed by using an ad hoc dynamic image analysis (DIA) device, based on the free falling particle scheme of particles per unit of volume. It resulted evident that nucleation, agglomeration, and breakage phenomena occur simultaneously during all the process time [47], phenomena that therefore must be taken into account in the modeling approach.

4. Mathematical approach to describe the phenomena involved in granulation process

Mathematical modeling of the granulation process can play a dual role: can help to understand and to underline the observed physical phenomena and to predict properties (size and size distribution) of granular materials. It is important to note that the predictive ability is a powerful way to define suitable process conditions, which can minimize costs and optimize process yields, avoiding onerous experimental tests.

Modeling approaches mainly consist in three types:

- Empirical models: they are obtained from the regression of large set of experimental data, most of the time derived from the design of experiments statistic technique [44, 48, 49]. Empirical models have the advantages of being very simple (often polynomial equations), easy to obtain (regression of data), quite reliable within the investigated range. The drawback of this approach is that it is a black box approach: it provides very little (or none) information on the underlying mechanisms.

- Discrete element method (DEM) models: this is the most detailed type of mathematical model for particulate systems [50]. Thanks to mass and momentum balances on each particle of the powder bulk, DEM models aim to describe the evolution of the analyzed system. The advantage is that the description is very detailed (at a single particle level), but it is so computational power demanding that rarely it is applied to systems of more than hundreds of thousand particles. They are often used to obtain parameters useful for higher scale models [51].

- Population balance equation (PBE)-based models: the powder bulk is described as a population with a distribution of certain characteristics (internal coordinates i.e., size, binder content) that can vary in space (external coordinates) and time [52]. Therefore, the PBE-based models describe the evolution of the number of particles of a given characteristic in time and space [53]. The advantages of such an approach, with respect to DEM models, are enormous in terms of computational power requirement, despite the complexity of the involved equations (integro-differential). Nowadays, this is still the most used approach in modeling granulation processes and several numerical methods have been developed to "easily" solve the PBEs. Among them, the discretization of the PBEs (DPBEs) in classes (multiclass method) is the most used to describe the evolution of the distribution of the internal variables (i.e., size: particle size distribution) [54–58]. Less computational demanding methods, like the method of moment (which focus only on the moments of distributions), are preferred when the PBE has to be combined with computational fluid dynamics (CFD) models to describe the multiphase flow (i.e., the movements inside the granulator) [59].

4.1. Discretized population balance equations modeling

4.1.1. From the continuous to the discretized form of PBEs

In the following, a modeling approach based on the one-dimensional (one internal coordinate: particle size) discretized PBEs (DPBEs) will be shown. Such kind of models turns to be the best choice when dealing with granulation processes, being very descriptive, since the evolution of the entire size distribution is considered, including all the phenomena that cause variation of the particle size distribution, and computational efficient (with the modern computational power), which allows to integrate them in flow sheet models of entire processes.

Considering the entire granulator as control volume, the PBEs do not depend on spatial variables (external coordinates), and it can be written as [60]:

$$
\frac{\partial n(v,t)}{\partial t} = \frac{\dot{Q}_{in}}{V} n_{in}(v) - \frac{\dot{Q}_{out}}{V} n_{out}(v) - \frac{\partial[(G-A)n(v,t)]}{\partial v}
$$
$$
+ B_{Nuc}(v,t) + B_{Agg}(v,t) - D_{Agg}(v,t) + B_{Br}(v,t) - D_{Br}(v,t) I.C.\ n(v,0) \tag{1}
$$
$$
= n_0(v) B.C.\ n(0,t) = 0
$$

where $n(v,t)$ is the probability density function $[b^{-1} x^{-1}]$, with "b" basis of calculation that can be the total mass of powder [kg] or the total volume $[m^3]$ and x the particle volume $[m^3]$ or particle diameter [m], depending on the internal coordinate chosen. Indeed, this last v [x] can be the particles volume v [m³] or the particles diameter l [m]. $\dot{Q}_{in}$ and $\dot{Q}_{out}$ are the flow rate of the inlet and outlet currents [b t⁻¹], V is the volume or mass in the granulator [b], and G and A are the growth as layering and the attrition [x t⁻¹], respectively. B_{Nuc} is the birth by nucleation [b⁻¹ x⁻¹ t⁻¹], which creates particles within the PSD of interest. B_{Agg} and D_{Agg} are the birth and death by agglomeration of particle of size v, and analogously B_{Br} and D_{Br} are the birth and death by breakage phenomena of particle of size v. Disregarding layering and the attrition phenomena, for a batch granulator:

$$\frac{\partial n(v,t)}{\partial t} = B_{Nuc}(v,t) + B_{Agg}(v,t) - D_{Agg}(v,t) + B_{Br}(v,t) - D_{Br}(v,t)$$

(2)

$$I.C. \ n(v,0) = n_0(v)$$

The nucleation, agglomeration, and the breakage phenomena are described by the equations:

$$B_{Nuc}(v,t) = K\delta(v)$$

(3)

$$B_{Agg}(v,t) = \frac{1}{2}\int_0^v \beta(u,v-u)n(u,t)n(v-u,t)du$$

(4)

$$D_{Agg}(v,t) = \int_0^\infty \beta(u,v)n(u,t)n(v,t)du$$

(5)

$$B_{Br}(v,t) = \int_v^\infty b(u,v)S(u)n(u,t)du$$

(6)

$$D_{Br}(v,t) = S(v)n(v,t)$$

(7)

where K is a constant that multiply a function of the internal coordinate v (i.e., Dirac delta function). β is the coalescence kernel [b t⁻¹], which describes the frequency of collision between particles of internal coordinate u and v and the influence of the internal coordinates on the efficiency of agglomeration. On the other hand, b is the breakage function [x⁻¹], which describes the probability of formation of particles of internal coordinates v from the collision and breakage of particles of internal coordinates u ($u > v$). S is a selection function [t⁻¹], which describes the frequency at which particles of a given internal coordinates are broken.

To solve Eq. (2), a numerical method has to be used: the class methods of zero order [61],

$$N_i = \int_{v_i'}^{v_{i+1}'} n(v)dv = n_i(v_{i+1}' - v_i')$$

(8)

where N_i is the number of particles per unit base [b⁻¹] within a class with size range $[v_i', v_{i+1}']$. The characteristic size of each class is v_i: $v_i' \le v_i \le v_{i+1}'$. With this approach, instead of solving for a continuous (particle size) distribution, a system of ordinary differential equations (ODEs) is generated to describe the evolution of the number of particles within the classes. Increasing the number of classes, the number of ODEs increases as well as the computational power requirements: the most used discretization uses a geometric progression with common ratio $v_{i+1}'/v_i' = 2^{1/q}$ (or equivalently $l_{i+1}'/l_i' = \sqrt[3q]{2}$), where q is an integer ≥ 1.

4.1.2. Modeling the agglomeration phenomena

The most used discretized form of the agglomeration phenomena is the one proposed by Hounslow et al. [55] ($v'_{i+1}/v'_i = 2$), updated by Litster et al. [62] to consider geometric progression with $q \geq 1$ ($v'_{i+1}/v'_i = 2^{1/q}$). The described agglomeration phenomenon is binary (interaction between two particles), and the birth terms are due to the collisions and coalescence between particles of lower dimensions with respect to the considered class. The death terms account for the interactions and coalescence of particles belonging to the considered class, producing their disappearance (and their birth in upper classes).

$$B_{Agg_i}(t) - D_{Agg_i}(t) =$$

$$\sum_{j=1}^{i-S(q)-1} \frac{2^{\frac{j-i+1}{q}}}{2^{\frac{1}{q}}-1} \beta(i-1,j)N_{i-1}N_j \; +$$

$$+ \sum_{k=2}^{q} \sum_{j=i-S(q-k+2)-k+1}^{i-S(q-k+1)-k} \frac{2^{\frac{j-i+1}{q}}-1+2^{\frac{-k+1}{q}}}{2^{\frac{1}{q}}-1} \beta(i-k,j)N_{i-1}N_j +$$

$$+ 0.5\beta(i-q,i-q)N_{i-q}^2 +$$

$$+ \sum_{k=2}^{q} \sum_{j=i-S(q-k+2)-k+2}^{i-S(q-k+1)-k+1} \frac{2^{\frac{j-i}{q}}-2^{\frac{-k+1}{q}}+2^{\frac{1}{q}}}{2^{\frac{1}{q}}-1} \beta(i-k+1,j)N_{i-k+1}N_j +$$

$$- \sum_{j=1}^{i-S(q)} \frac{2^{\frac{j-i}{q}}}{2^{\frac{1}{q}}-1} \beta(i,j)N_iN_j \; - \sum_{j=1-S(q)+1}^{h} \beta(i,j)N_iN_j$$

(9)

where $S(q) = \sum_{p=1}^{q} p$. For $q = 1$, the Hounslow discretization can be obtained.

4.1.2.1. The kernel of coalescence

The coalescence kernel is the most important parameter when describing granulation processes. A body of literature is present on this kernel, proposing several expressions ranging from purely empirical, semiempirical, and model-based kernels. In general, the coalescence kernel is split into two parts:

$$\beta(u,v,t) = \beta_0(t)\beta(u,v) \tag{10}$$

where the first is the "aggregation rate" term and the latter describes the dependence of the coalescence kernel on the dimensions of the granules. In the first term, various system parameters are incorporated (i.e., granulator geometry, operating conditions, formulation properties, etc.).

The nature of $\beta(u,v)$, most of the time an homogeneous function, establishes how the agglomeration modify the internal coordinate, in particular whether or not the transformation is self-preserving in PSD and whether or not a gelling behavior should be expected. Analyzing the degree of homogeneity λ $(\beta(cu, cv) = c^\lambda \beta(u, v))$, which expresses the strength of the dependence of $\beta(u, v)$ on its argument, it is possible to distinguish between nongelling and leading to a self-preserving size distribution kernels ($\lambda \leq 1$) and gelling (and non-self-preserving PSD) kernels ($\lambda > 1$) [46]. *The most used kernel in literature is the equikinetic energy (EKE) kernel, which is a nongelling* and leading to a self-preserving size distribution kernels, as it can be seen from **Table 1**. In **Figure 3**, the (normalized) EKE kernel is reported: as it can be seen the maximum probability of coalescence in a binary process is between a big particle (high classes) and a small particle (low classes).

4.1.3. Modeling the breakage phenomena

The discretized form of the breakage phenomena can be obtained by substituting continuous with discrete functions and by using Eq. (8):

$$B_{Bri}(t) - D_{Bri}(t) = \sum_{j=i+1}^{\infty} b(v_i, v_j) S(v_j) N_j \Delta v_j' - S_i N_i \tag{11}$$

However, as suggested in Vanni [63], to satisfy the mass conservation (valid for both agglomeration an breakage), for all the possible discretizations (i.e., for a geometric progression of the type $v_{i+1}'/v_i' = 2^{1/q}$), the equation has to be corrected:

$$B_{Bri}(t) - D_{Bri}(t) = \sum_{j=i+1}^{\infty} \Gamma_{i,j} S_j N_j C_j^{(1)} - C_i^{(2)} S_i N_i \tag{12}$$

where

$$C_i^{(1)} = \frac{v_i}{\sum\limits_{j=1}^{i-1} v_j \Gamma_{ji}} C_i^{(2)} \tag{13}$$

$$C_i^{(2)} = 1 - \frac{1}{v_{i+1}' - v_i'} \int_{v_i'}^{v_{i+1}'} \left[\int_v^{v_{i+1}'} b(v, q) dq \right] dv \tag{14}$$

Kernel	Equation	λ
Constant kernel	$\beta(u, v) = 1$	0
Sum kernel	$\beta(u, v) = u + v$	1
Product kernel	$\beta(u, v) = u\,v$	2
Coagulation kernel	$\beta(u, v) = u^{\frac{1}{3}} + v^{\frac{1}{3}}$	2/3
Equikinetic energy kernel	$\beta(u, v) = \left(u^{\frac{1}{3}} + v^{\frac{1}{3}}\right)^2 \sqrt{\frac{1}{u} + \frac{1}{v}}$	1/6

Table 1. Examples of coalescence kernels and degrees of homogeneity.

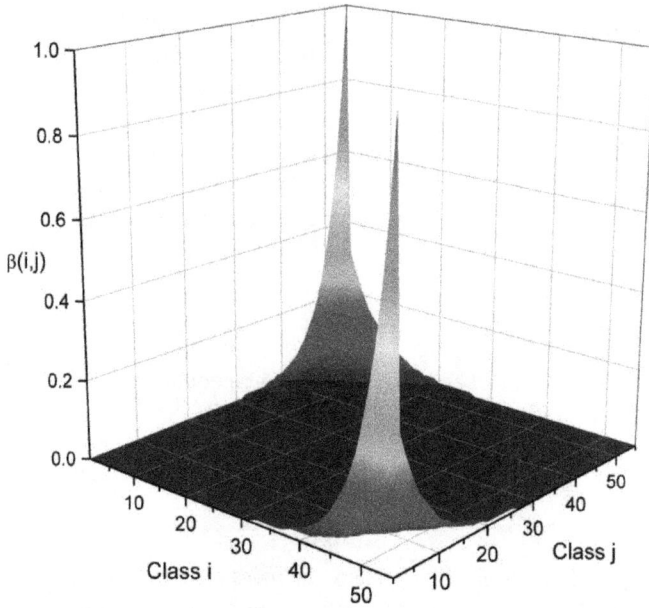

Figure 3. Equikinetic energy coalescence kernel.

$$\Gamma_{ij} = \int_{v_i'}^{v_{i+1}'} b(v_i, v_j) dv \qquad (15)$$

with $C_1^{(2)} = 0$ and v_i the characteristic size x of the class in volume [m³]. In case the distribution density function has been obtained $(n(l, t))$ and discretized $(N_i = n_i(l_{i+1}' - l_i'))$ with the diameter l_i as caracteristic size, the relation $v_i \sim l_i^3$ can be used to adapt Eqs. (13), (14), and (15).

4.1.3.1. The selection and breakage function

Several functions can be chosen for the selection function and breakage function (continuous or discrete), usually of semiempirical nature: the breakage theory is not well developed as the agglomeration theory [63].

An example of selection function can be the power law form: $S_i = kv_i^\gamma$, where k and γ $(1/3 \leq \gamma \leq 2)$ are adjustable parameters.

A flexible form of the breakage function is the parabolic form:

$$b(v_i, v_j) = \frac{C}{v_j - 1} + \left(1 - \frac{C}{2}\right)\left[\frac{8\,(3v_i^2 - 3v_i + 1)}{(v_j - 1)^3} - \frac{12(2v_i - 1)}{(v_j - 1)^2} + \frac{6}{v_j - 1}\right] \qquad (16)$$

that, depending on C, can simulate different behaviors: concave parabola $(0 \leq C < 2)$, it is more likely the formation of unequal fragments, convex parabola $(2 < C < 3)$, it is more likely the formation of equal fragments, and uniform distribution ($C = 2$), in which it is equally likely to form a child particle of any size.

4.1.4. Modeling the nucleation phenomenon

Nucleation phenomenon occurs because small particles lower than the considered minimum size class can suddenly form granules (i.e., due to the action of binder droplets) within the considered size range. In light of this, it is clear that nucleation is not a mass conservative mechanism. It can be modeled as:

$$B_{Nuc_i}(t) = k(t)f(i) \tag{17}$$

where $k(t)$ is a function of time, describing for how long this phenomenon is present. $f(i)$ is a function of the size class and it individuates the class interested by nucleation.

4.1.5. Model results

The resulting DPBE is a system of ordinary differential equations:

$$\frac{dN_i}{dt} = B_{Agg_i}(t) - D_{Agg_i}(t) + B_{Bri}(t) - D_{Bri}(t) + B_{Nuc_i}(t) \tag{18}$$
$$I.C. N_i(0) = N_{i0},$$

which can be solved numerically with the well-known discretization techniques (i.e., explicit Runge-Kutta). The results give the evolution of N_i during the granulation process, allowing to follow the (discretized) particles size distribution (**Figure 4**).

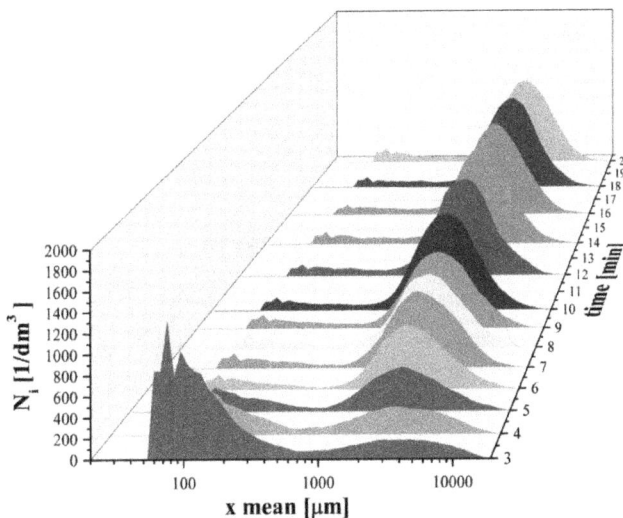

Figure 4. Time evolution of the PSD in terms of number of particle per unit volume N_i (impeller rotation speed of 93 rpm and liquid binder flow rate of 17 ml/min).

The agglomeration process leads to the lowering and translation toward higher dimension of the PSD, because the particles diminish in number increasing their size. On the contrary, breakage phenomena lead to an increase of the number of particles per unit volume, with a translation toward lower dimensions of the PSD. Finally, nucleation locally increases the number of particles per unit volume.

5. Conclusions

Granular materials represent a relevant form of commercial products on the worldwide market. A rational organization of manufacturing, based on phenomenological knowledge rather than practical trials, can allow high granulation performance in terms of yields, product features, and manufacturing costs.

In this chapter, wet granulation process is described presenting the role of several factors, such as apparatus features, formulation, and operative parameters on granules final features.

Experimental campaigns of wet granulation runs, performed by using a low-shear granulator, changing several formulation and process parameters, have been performed to detail each single parameter effect both on the granulation stages and on the obtained granular materials features.

Mathematical modeling approaches of the granulation process have been thus introduced with the aim both to understand and underline the observed physical phenomena, and to propose predictive tools able to forecast granulation results in terms of size and size distribution of obtained granular materials. The predictive ability is a powerful way to define suitable process conditions, which can minimize costs and optimize process yields, avoiding onerous experimental tests. In this work, particular attention was given to models based on population balance equations (PBEs), for which appropriate mathematical descriptive functions have been presented.

Conflict of interest

The authors declare that they have no conflict of interests.

Abbreviations

ASTM	American society for testing and materials
CCD	central composite design
CFD	computational fluid dynamics

DEM	discrete element method
DIA	dynamic image analysis
DoE	design of experiments
DPBE	discretized population balance equation
EKE	equikinetic energy
HPMC	hydroxypropyl methylcellulose
ODE	ordinary differential equation
PBE	population balance equation
PSD	particle size distribution

Symbols

β	coalescence kernel
A	attrition
b	breakage function
B_{Agg}	birth by agglomeration phenomena of particle of size v
B_{Nuc}	birth by nucleation phenomena
B_{Br}	birth by breakage phenomena of particle of size v
D_{Br}	death by breakage phenomena of particle of size v
D_{Agg}	death by agglomeration phenomena of particle of size v
G	growth as layering
K	constant that multiply a function of the internal coordinate v
n	probability density function
N_i	number of particles for class "i" per unit of volume of solid
q	integer ≥ 1
$\dot{Q}_{in}$	flow rate of the inlet currents
$\dot{Q}_{out}$	flow rate of the outlet currents
S	selection function
v	internal coordinate (particle volume or particle diameter)
V	volume or mass in the granulator

Author details

Veronica De Simone[1,2], Diego Caccavo[1], Annalisa Dalmoro[2], Gaetano Lamberti[1], Matteo d'Amore[2] and Anna Angela Barba[2]*

*Address all correspondence to: aabarba@unisa.it

1 Department of Industrial Engineering, University of Salerno, Fisciano, SA, Italy

2 Department of Pharmacy, University of Salerno, Fisciano, SA, Italy

References

[1] Perry RH, Green DW. Perry's Chemical Engineers' Handbook. USA: McGraw-Hill Professional; 1999

[2] Iveson SM, Litster JD, Hapgood K, Ennis BJ. Nucleation, growth and breakage phenomena in agitated wet granulation processes: A review. Powder Technology. 2001;**117**(1):3-39

[3] Hapgood KP, Litster JD, Smith R. Nucleation regime map for liquid bound granules. American Institute of Chemical Engineers AIChE Journal. 2003;**49**(2):350

[4] Wright T. Solid Dosage Outsourcing Trends. NJ, USA: Contract Pharma Rodman Media; 2016

[5] Langhauser K. Long Live OSD. In: Solid Dose Trends [Internet]. Pharmaceutical Manufacturing [21]. 2017. Available from: https://www.pharmamanufacturing.com/

[6] Van Arnum P. Evaluating Market Opportunities for Solid Dosage Products and Manufacturing. 2015. Available from: https://www.dcatvci.org/1

[7] FAO. World Fertilizer Trends and Outlook to 2020. 2017

[8] Research GV. Animal Feed Additives Market Analysis By Product (Antioxidants, Amino Acids), Enzymes (Phytase, Non-Starch Polysaccharides), Acidifiers, By Livestock (Pork/ Swine, Poultry, Cattle, Aquaculture), And Segment Forecasts, 2018–2025. CA, USA: Grand View Research; 2017

[9] Technology SG. Available from: www.steecker.com

[10] Iveson S, Litster J. Growth regime map for liquid-bound granules. AICHE Journal. 1998; **44**(7):1510-1518

[11] Mahdi F, Hassanpour A, Muller F. An investigation on the evolution of granule formation by in-process sampling of a high shear granulator. Chemical Engineering Research and Design. 2018;**129**:403-411

[12] Luo G, Xu B, Zhang Y, Cui X, Li J, Shi X, et al. Scale-up of a high shear wet granulation process using a nucleation regime map approach. Particuology. 2017;**31**:87-94

[13] Suresh P, Sreedhar I, Vaidhiswaran R, Venugopal A. A comprehensive review on process and engineering aspects of pharmaceutical wet granulation. Chemical Engineering Journal. 2017;**328**:785-815

[14] Shanmugam S. Granulation techniques and technologies: Recent progresses. BioImpacts: BI. 2015;**5**(1):55

[15] Nalesso S, Codemo C, Franceschinis E, Realdon N, Artoni R, Santomaso AC. Texture analysis as a tool to study the kinetics of wet agglomeration processes. International Journal of Pharmaceutics. 2015;**485**(1):61-69

[16] Parikh DM. Handbook of Pharmaceutical Granulation Technology. FL, USA: CRC Press; 2016

[17] Litster J, Ennis B. The Science and Engineering of Granulation Processes. Netherlands: Springer Science & Business Media; 2013

[18] Morin G, Briens L. A comparison of granules produced by high-shear and fluidized-bed granulation methods. AAPS PharmSciTech. 2014;**15**(4):1039-1048

[19] Guigon P, Simon O, Saleh K, Bindhumadhavan G, Adams M, Seville J. Handbook of Powder Technology. Netherlands: Elsevier Science BV; 2007. pp. 255-288

[20] Verstraeten M, Van Hauwermeiren D, Lee K, Turnbull N, Wilsdon D, Ende M, et al. In-depth experimental analysis of pharmaceutical twin-screw wet granulation in view of detailed process understanding. International Journal of Pharmaceutics. 2017;**529**(1):678-693

[21] Sarkar S, Chaudhuri B. DEM modeling of high shear wet granulation of a simple system. Asian Journal of Pharmaceutical Sciences. 2018;**13**(3):220-228

[22] Mirza Z, Liu J, Glocheux Y, Albadarin AB, Walker GM, Mangwandi C. Effect of impeller design on homogeneity, size and strength of pharmaceutical granules produced by high-shear wet granulation. Particuology. 2015;**23**:31-39

[23] Walker G, Andrews G, Jones D. Effect of process parameters on the melt granulation of pharmaceutical powders. Powder Technology. 2006;**165**(3):161-166

[24] Knight P, Instone T, Pearson J, Hounslow M. An investigation into the kinetics of liquid distribution and growth in high shear mixer agglomeration. Powder Technology. 1998; **97**(3):246-257

[25] Morkhade DM. Comparative impact of different binder addition methods, binders and diluents on resulting granule and tablet attributes via high shear wet granulation. Powder Technology. 2017;**320**:114-124

[26] Oka S, Smrčka D, Kataria A, Emady H, Muzzio F, Štěpánek F, et al. Analysis of the origins of content non-uniformity in high-shear wet granulation. International Journal of Pharmaceutics. 2017;**528**(1):578-585

[27] Osborne JD, Sochon RPJ, Cartwright JJ, Doughty DG, Hounslow MJ, Salman AD. Binder addition methods and binder distribution in high shear and fluidised bed granulation. Chemical Engineering Research and Design. 2011;**89**(5):553-559

[28] Sakr WF, Ibrahim MA, Alanazi FK, Sakr AA. Upgrading wet granulation monitoring from hand squeeze test to mixing torque rheometry. Saudi Pharmaceutical Journal. 2012;**20**(1): 9-19

[29] Iveson S, Litster J, Ennis B. Fundamental studies of granule consolidation Part 1: Effects of binder content and binder viscosity. Powder Technology. 1996;**88**(1):15-20

[30] Iveson S, Litster J. Fundamental studies of granule consolidation Part 2: Quantifying the effects of particle and binder properties. Powder Technology. 1998;**99**(3):243-250

[31] Bertín D, Cotabarren IM, Veliz Moraga S, Piña J, Bucalá V. The effect of binder concentration in fluidized-bed granulation: Transition between wet and melt granulation. Chemical Engineering Research and Design. 2018;**132**:162-169

[32] Hoornaert F, Wauters PA, Meesters GM, Pratsinis SE, Scarlett B. Agglomeration behaviour of powders in a Lödige mixer granulator. Powder Technology. 1998;**96**(2):116-128

[33] Schæfer T, Johnsen D, Johansen A. Effects of powder particle size and binder viscosity on intergranular and intragranular particle size heterogeneity during high shear granulation. European Journal of Pharmaceutical Sciences. 2004;**21**(4):525-531

[34] Ennis B, Litster J. Particle size enlargement. In: Perry's Chemical Engineers' Handbook. 7th ed. New York: McGraw-Hill; 1997. p. 20(20.89)

[35] Seo A, Holm P, Schæfer T. Effects of droplet size and type of binder on the agglomerate growth mechanisms by melt agglomeration in a fluidised bed. European Journal of Pharmaceutical Sciences. 2002;**16**(3):95-105

[36] Fan X, Yang Z, Parker DJ, Ng B, Ding Y, Ghadiri M. Impact of surface tension and viscosity on solids motion in a conical high shear mixer granulator. AICHE Journal. 2009; **55**(12):3088-3098

[37] Schaefer T, Holm P, Kristensen H. Melt granulation in a laboratory scale high shear mixer. Drug Development and Industrial Pharmacy. 1990;**16**(8):1249-1277

[38] Bellocq B, Cuq B, Ruiz T, Duri A, Cronin K, Ring D. Impact of fluidized bed granulation on structure and functional properties of the agglomerates based on the durum wheat semolina. Innovative Food Science & Emerging Technologies. 2018;**45**:73-83

[39] Badawy SIF, Hussain MA. Effect of starting material particle size on its agglomeration behavior in high shear wet granulation. AAPS PharmSciTech. 2004;**5**(3):16-22

[40] Holm P. Effect of impeller and chopper design on granulation in a high speed mixer. Drug Development and Industrial Pharmacy. 1987;**13**(9–11):1675-1701

[41] Bareschino P, Marzocchella A, Salatino P. Fluidised bed drying of powdered materials: Effects of operating conditions. Powder Technology. 2017;**308**:158-164

[42] Srivastava S, Mishra G. Fluid bed technology: Overview and parameters for process selection. International Journal of Pharmaceutical Sciences and Drug Research. 2010;**2**(4): 236-246

[43] Badawy SIF, Menning MM, Gorko MA, Gilbert DL. Effect of process parameters on compressibility of granulation manufactured in a high-shear mixer. International Journal of Pharmaceutics. 2000;**198**(1):51-61

[44] De Simone V, Dalmoro A, Lamberti G, d'Amore M, Barba AA. Central composite design in HPMC granulation and correlations between product properties and process parameters. New Journal of Chemistry. 2017;**41**(14):6504-6513

[45] De Simone V, Dalmoro A, Lamberti G, Caccavo D, d'Amore M, Barba AA. HPMC granules by wet granulation process: Effect of vitamin load on physicochemical, mechanical and release properties. Carbohydrate Polymers. 2018;**181**:939-947

[46] De Simone V, Dalmoro A, Lamberti G, Caccavo D, d'Amore M, Barba AA. Effect of binder phase and load solubility properties on HPMC granules produced by wet granulation process. Carbohydrate Polymers. 2018 (submitted)

[47] De Simone V, Caccavo D, Lamberti G, d'Amore M, Barba AA. Wet-granulation process: Phenomenological analysis and process parameters optimization. Powder Technology. 2018 (submitted)

[48] Mangwandi C, Albadarin AB, AAH A-M, Allen SJ, Walker GM. Optimisation of high shear granulation of multicomponent fertiliser using response surface methodology. Powder Technology. 2013;**238**:142-150

[49] Aleksić I, Đuriš J, Ilić I, Ibrić S, Parojčić J, Srčič S. In silico modeling of in situ fluidized bed melt granulation. International Journal of Pharmaceutics. 2014;**466**(1):21-30

[50] Sen M, Barrasso D, Singh R, Ramachandran R. A multi-scale hybrid CFD-DEM-PBM description of a fluid-bed granulation process. PRO. 2014;**2**(1):89

[51] Lee KF, Dosta M, McGuire AD, Mosbach S, Wagner W, Heinrich S, et al. Development of a multi-compartment population balance model for high-shear wet granulation with discrete element method. Computers and Chemical Engineering. 2017;**99**:171-184

[52] Ramkrishna D. Population Balances: Theory and Applications to Particulate Systems in Engineering. USA: Elsevier Science; 2000

[53] Abberger T. Population balance modelling of granulation Chapter 24. In: Salman AD, Hounslow MJ, Seville JPK, editors. Handbook of Powder Technology. Vol. 11. Netherlands: Elsevier Science B.V; 2007. pp. 1109-1186

[54] Yang C, Mao Z-S. Chapter 6—Crystallizers: CFD–PBE modeling. In: Numerical Simulation of Multiphase Reactors with Continuous Liquid Phase. Academic Press: Oxford; 2014. pp. 263-294

[55] HM J, RR L, MV R. A discretized population balance for nucleation, growth, and aggregation. AICHE Journal. 1988;**34**(11):1821-1832

[56] Kumar S, Ramkrishna D. On the solution of population balance equations by discretization—I. A fixed pivot technique. Chemical Engineering Science. 1996;**51**(8):1311-1332

[57] Kumar S, Ramkrishna D. On the solution of population balance equations by discretization—II. A moving pivot technique. Chemical Engineering Science. 1996;**51**(8):1333-1342

[58] Kumar S, Ramkrishna D. On the solution of population balance equations by discretization—III. Nucleation, growth and aggregation of particles. Chemical Engineering Science. 1997;**52**(24):4659-4679

[59] Marchisio DL, Fox RO. Computational Models for Polydisperse Particulate and Multiphase Systems. New York, USA: Cambridge University Press; 2013

[60] Litster J, Ennis B. The Science and Engineering of Granulation Processes. Netherlands: Springer; 2004

[61] Salman AD, Hounslow M, Seville JPK. Granulation. Netherlands: Elsevier Science; 2006

[62] LJ D, SD J, HM J. Adjustable discretized population balance for growth and aggregation. AICHE Journal. 1995;**41**(3):591-603

[63] Vanni M. Approximate population balance equations for aggregation-breakage processes. Journal of Colloid and Interface Science. 2000;**221**(2):143-160

www.ingramcontent.com/pod-product-compliance
Lightning Source LLC
Chambersburg PA
CBHW081236190326
41458CB00016B/5804